STUDIES IN ZOOLOGY

STUDIES IN ZOOLOGY

(For B.Sc I & II)

Dr. P. Raikumar Singh

AKANSHA PUBLISHING HOUSE
NEW DELHI-110002 (INDIA)

AKANSHA PUBLISHING HOUSE
4649-B/21, Ansari Road
Darya Ganj, New Delhi - 110 002
Ph.: 23263193 / 9811582579 / 09435116718
Fax: 011-23263193
Email: ektabooks@yahoo.com

Studies in Zoology (For B.Sc I & II)

First Published 2009
ISBN 978-81-8370-172-3

[The responsibility for the facts stated, conclusions reached etc., is entirely that of the author. The publisher is not responsible for them, whatsoever.]

PRINTED IN INDIA

Published by Akansha Publishing House, New Delhi and Printed at Aryan Enterprises, Delhi.

CONTENTS

Foreword vii

1 Phylum Protozoa 3

2 Phylum Platyhelminthes 48

3 Phylum Nemathelminthes 78

4 Phylum Annelida 91

5 Phylum Arthropoda 124

6 Nervous System of Mollusca 153

7 Phylum Echinodermata 165

8 Accounts of Minor Phyla 184

Questions & Answers 213

FOREWORD

This book is my humble presentation as a tribute to my late husband Dr. P. Raikumar Singh who has authored this book during his mundane existence. He was an academician and a dedicated teacher whose sole aim in life was the welfare of the students. His love and concern for his students was so intense that he could write six Textbooks and guides in his favourite subject Zoology in spite of his busy schedule.

After his retirement from govt. service in 1996, he wrote a book entitled 'Studies in Zoology," keeping in view the needs of B.Sc. students. But unfortunately, he died before the book could be published. He left the manuscript in his own beautiful hand writing and the same have been printed without any addition/deletion/modification.

Zoology is a very significant discipline of study in understanding human nature. Its applications find it way into medical practices and for the improvement of health and care of patients.

Researches in the field of Zoology is now being conducted widely and there is no dearth of books and research papers on this subject. And in such a present scenario or simple book like the present one would hardly matter. Yet I sincerely hope that the book will be useful to students of undergraduate standard. I am aware of the shortcomings of the publication as the book is being produced when the author is no more. I hope and trust that all the omissions and commissions in this publication may be excused.

I would be failing in my duties unless I record my deep sense of gratitude to Dr. Ayekpam Chinglenkhomba Meetei for his untiring work in reading the manuscript and making improvement in the book with appropriate diagrams which he did as a labour of love for his teacher-oja Raikumar.

Jamini Devi
Chairperson
Manipur State Women's Commission
Imphal/Manipur

Non-Chordata
(INVERTEBRATA)

Invertebrates

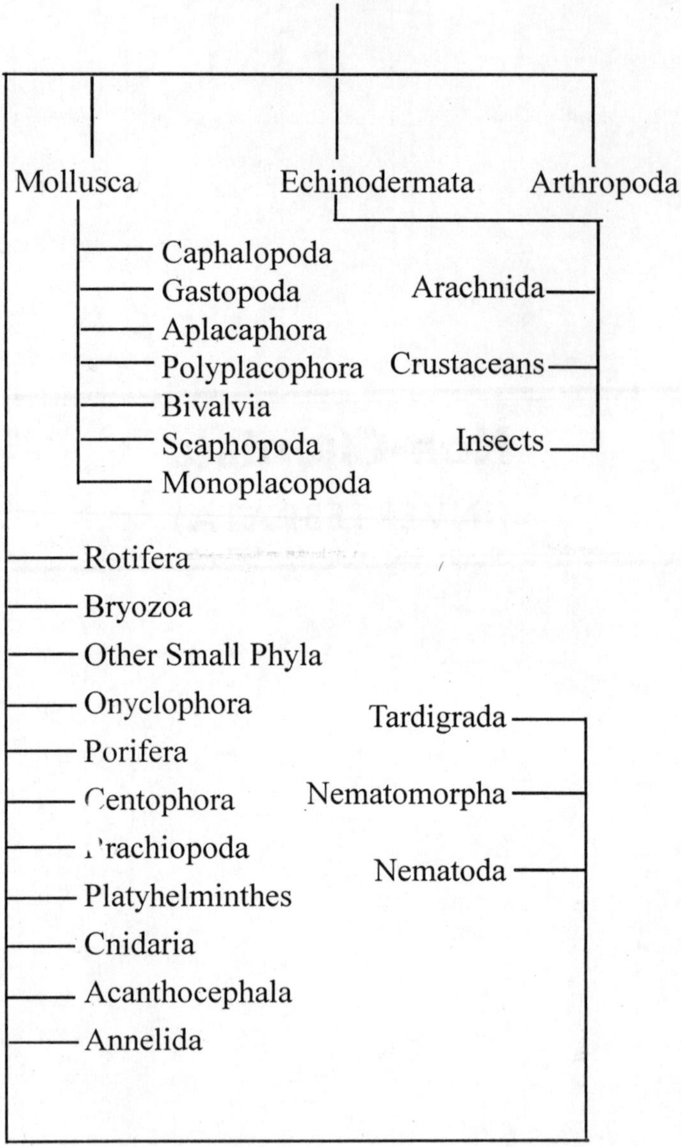

1
PHYLUM PROTOZOA

Distinguishing Characters of Protozoa

The Protozoa are unicellular animals without tissues or organs existing singly or in colonies of a few to thousands of individuals; such colonies differ from Metazoa in that, their components are all alike except when engaged in reproductive activity.

The distinctive characters of the Phylum Protozoa are: –

1. They are all small and microscopic (measuring in microns) with a single cell; some forming colonies ranging from few to many similar individuals; no symmetry, bilateral, radial or spherical.

2. Cell form usually constant, oval, elongate, spherical or otherwise.

3. Distinct nucleus, may be single or multiple; other body parts are in the form of organelles ; but no organs or tissue systems.

4. Some species form protective housings or tests (shell), many others produce resistant cysts, etc. to face unfavourable circumstances.
5. Locomotion is effected by the structural parts of the body namely, pseudopodia, flagella, cilia, or by the whole body itself.
6. Mode of existence, free-living, commensalistic, parasitic or otherwises.
7. Nutrition is effected in many ways: (a) Holophytic, by means of photosynthesis, (b) Saprophytic, subsisting on dissolved substances in their surroundings, (c) Holozoic, subsisting on other organisms, or (d) Parasitic.
8. Excretion through general surface or through contractile vacuoles which also serve for osmoregulation.
9. Respiration by the body itself.
10. Reproduction by (a) asexual, produced by binary fission, multiple fission or budding, and (b) sexual, produced by the fusion of the gametes or conjugation.
11. Encystment commonly occurs to help in dispersal and to resist unfavourable conditions.
12. Many forms with complicated life histories, like alternation of generations, etc.

Classification of Protozoa

The classification of the Phylum Protozoa may be based on the (a) nutrition of these animals – Protozoa, and (b) on the locomotory organs of these animals – Protozoa, or (c) on the combination of the above two methods.

The classification based on the locomotor organs of these animals – Protozoa, are much appreciating and widely followed by many authors. And considering this, we will also divide the Phylum Protozoa – into 5 (five) classes depending on the nature of the "locomotor organs". (Parker and Haswell)

Class I. Rhizopoda (Sarcodina)

Characters

1. The class Rhizopoda or Sarcodina (Gr. *Rhizos* = root, *podos* = feet, *Sarcodes* = fleshy) are naked and shell bearing Protozoa in which the body differentiation is not reached to an extreme degree.

2. Practically they are devoid of any symmetry, or some may carry spherical symmetry; where there is no spherical symmetry, the body form keeps constantly changing; while again, in other shell bearing forms the shape of the body depends on the shell. The nature of the shell varies from simple to complex many chambered.

3. The cell body is divided into ecto and endoplasm; ectoplasm lends form to the cell body, endoplasm comprises many other inclusions, and the nucleus, which occupies no fixed position. Nucleus may be one to many in the body : it is usually 'vesicular' type containing a larger quantity of nuclear sap, which is relatively more fluid and surrounds one or more intranuclear bodies called 'endosomes' ; the endosomes may consist of chromatin alone when it is known as 'karyosome', or it may consist of plastin alone when it is known as 'nucleolus' or 'plasmosome'.

4. Members of the class perform their movement by the formation of a finger like extrusions of the cell-body-ectoplasm, known as pseudopodia. These pseudopodia may be (i) blunt, constantly changing lobopods (e.g. Amoeba), (ii) long, fine thread-like filopods (e.g. Heliozoa), or anastomosing reticulate (e.g. Foraminifera), etc.

5. Nutrition is of holozoic mode.

6. Respiration and excretion mainly by the cell body itself.
7. Reproduction is carried out mostly in two ways : (a) asexual, by binary fission or a division of the body into two, or multiple fission, or budding, etc., and (b) sexual.
8. Life cycle comprises amoeboid and flagellate stages.
9. They are solitary, free living, parasitic, or found in fresh and salt waters.

The class Rhizopoda (or Sarcodina) is divided into 5 (five) orders according to Parker and Haswell.

Order 1. Lobosa (or Amoebina or Amoebozoa)

(i) Fresh water rhizopods with blunt and lobose pseudopodia.

(ii) May or may not present a shell or skeleton in them.

Examples : (1) *Amoeba proteus*

(2) *Entamoeba hystolytica*

(3) *E. gingivalis.*

Order 2. Filosa

(i) Pseudopodia fine and thread-like and becoming branched towards the ends. Example : (1) *Euglypha.*

Order 3. Foraminifera

(i) Marine rhizopods having pseudopodia which are slender and branched, the branches interlacing to form a network.

(ii) A calcareous shell is present in them, which may be simple or many chambered.

(iii) Reproduction is by multiple fission, involving an alternation of generations. Examples: (1) *Polystomella* (or *Elphidium*) (2) *Allogromia*, (3) *Globigerina*, etc.

Order 4. Heliozoa

(i) Predominately fresh water rhizopods having ray-like pseudopodia extruded from the central region ; the pseudopodium is further supported by a fine and slender needle like structure known as axial filament (or Axoneme).

(ii) Food vacuoles and contractile vacuoles are present.

(iii) Nutrition is holozoic.

(iv) Uninucleate forms reproduce asexually by binary fission. Examples : (1) *Actinophrys* sol, (2) *Clathrulina*, etc.

Order 5. Radiolaria

(i) They live at great depths or at the surface of the marine waters, having mostly spherical in form.

(ii) Pseudopodia ray-like, stiff with a central filament.

(iii) Protoplasm of the body separated into inner and outer parts by porous spherical capsule of chitin.

(iv) Usually with outer skeleton of silica or strontium sulphate, with radiating spines.

(v) Nutrition is holozoic and reproduction is by binary fission. Example: (1) *Acanthometron*, (2) *Lithocercus*, etc.

Class II. Mastigophora (or Flagellata)

1. Body covered by a thin pellicle or test of cellulose, chitin or silica.

2. Presence of one or more flagella for locomotion and food capture; some with pseudopodia. Flagella are whip-like structures arising from a stalk of the body known as blepharoplast.

3. Some members exhibit both amoeboid and flagellate stages which indicate a close relationship of the group with Sarcodina (Rhizopoda).
4. Many of them possess chlorophyll (green pigment).
5. Nutrition-autotrophic, heterotrophic or mixotrophic.
6. Reproduction – asexual reproduction by longitudinal binary fission or sporulation.
7. Free living or parasitic.

Order 1. Chrysomonadina

(i) Small amoeboid forms with one or two flagella.

(ii) No gullet in them but having brown or yellow chromatophores.

(iii) Nutrition holophytic, sometimes saprophytic or holozoic.

(iv) Fresh and marine waters in distribution. Examples: (1) *Chrysamoeba*, (2) *Synura*, (3) *Ochromonas*, etc.

Order 2. Cryptomonadina

(i) Small, oval flagellates of constant form with rigid pellicle.

(ii) Flagella one or two, sometimes three.

(iii) Rarely amoeboid with gullet.

(iv) Nutrition holophytic or saprophytic.

(v) In fresh and marine waters. Examples:

(1) *Cryptomanas* (coloured),

(2) *Chilomonas* (colourless).

Order 3. Euglenoidea (Euglenoidina)

(i) Large in form, with a thick pellicle.

(ii) One or two flagella, mouth aperture, gullet, contractile vacuole, green chromatophores, stigma present.

(iii) Nutrition holophytic, holozoic or saprophytic.

(iv) Mostly inhabit fresh and stagnant waters.

Examples : (1) *Euglena viridis* , (2) *Copromonas*, etc.

Order 4. Phytomonadina (Volvocina)

(i) Small, solitary or colonial, with a rigid cellulose covering.

(ii) Green with chlorophyll, and have an eye spot.

(iii) Flagella two or four, sometimes more.

(iv) No cytostome or gullet.

(v) Nutrition holophytic. Examples: (1) *Volvox*, (2) *Chlamydomonas*.

Order 5. Dinoflagellata

(i) Body small, planktonic, naked or encased by cellulose plates.

(ii) A large non-contractile sac and a system of canals ; with or without gullet.

(iii) Two flagella – one directing forward and the other encircling the body.

(iv) Chromatophores many or none, with varying colours.

(v) Holophytic in nutrition.

(vi) Reproduction by fission.

Example: (1) *Ceratium*, (2) *Blastodinium*, etc.

Order 6. Cystoflagellata

(i) Body form spherical or otherwise.

(ii) Presence of a long flagellum and a short tentacle.

(iii) Holozoic in nutrition.

(iv) Mostly in marine waters. Example : (1) *Noctiluca.*

Order 7. Protomonadina

(i) Body small, somewhat amoeboid, free-living or parasitic. Colourless body without cuticle, mouth or gullet.

(ii) Flagella one to three.

(iii) Kinetoplast present or secondarily absent.

(iv) Nutrition is holozoic in free-living forms and saprophytic in parasitic forms.

(v) Reproduction by longitudinal fission ; multiple fission in some parasitic forms.

Examples : (1) *Trypanosoma* (causing sleeping sickness),

2) *Leishmania* (causing kala-azar in India).

Order 8. Polymastigina

(i) Body form oval and elongated, with gullet.

(ii) With 3 to 8 flagella, axostyle present.

(iii) Slit like or elongated mouth may be present.

(iv) Nutrition – holozoic or saprophytic.

(v) Reproduction – by longitudinal fission ; cyst formation commonly takes place to help in the transmission of parasitic forms.

(vi) They are most commensal or parasitic in both invertebrates and vertebrates.

Examples : (1) *Trichomonae vaginalis*,

(2) *Giardia enterica* (in intestine of man and vertebrates).

Class III. Sporozoa

1. The sporozoa (Gr. *Spora*=Seed + *Zoon*=Animal) are practically all parasitic Protozoa – of intercellular or intracellular in kind, occurring in invertebrates and vertebrates.

2. In the vegetative (Trophozoite i.e., feeding and growing) phases of the organism, the cell body assumes a rounded or elongated form like a worm, with one nucleus, having no locomotor organelles or contractile vacuoles; the nucleus may be of vesicular type whose shape is usually oval and contains a large quantity of nuclear sap of relatively more fluid encircling one or more intranuclear bodies called "endosomes".

3. Most Sporozoans move by change in shape of the cell body (gliding movements).

4. The body cytoplasm may consist of ecto and endoplasm; in the deeper part of the ectoplasm (in some cases, e.g., Monocystis) are present contractile elements called "Myonemes", which may cause the contraction of the whole or part of the body; some may possess special organelles for adhesion.

5. Nutrition is saprozoic (food matter is absorbed directly from the host).

6. Excretion and respiration are effected by simple diffusion of the cell body.

7. Reproduction – asexual by multiple fission, and sexual by gamete formation ; there may be flagella or amoeboid stages in the life history.

Sub Class A. Telosporidia

1. The trophic phase (trophozoite) is quite distinct and uninucleoted.
2. With or without polar capsules in spores ; spores hatch out as young trophozoites.

Order 1. Gregarinida

(i) Trophozoite – elongated and worm-like, 10 m to 16 mm. long.

(ii) Zygote – non-motile, producing one-walled spores, containing 8 sporozoites.

(iii) They are parasites of invertebrates ; they may be found in the gut, the colon or in the tissues.

Examples : (1) *Monocystis*, (2) *Gregarina*.

Order 2. Coccidia

(i) Mature trophozoite minute and intracellular.

(ii) Zygote non-motile.

(iii) Spores with one to many walls.

(iv) With alternate asexual schizogony followed by sporogony.

(v) With micro and mega-gametes.

(vi) They are intracellular parasites in epithelial tissues.

Examples : (1) *Eimerea* (*Coccidium*),
(2) *Isopora*, (3) *Lankestrella*, etc.

Order 3. Haemosporidia

(i) They are intracellular parasites in the RBC or the reticulo endothelial cells of vertebrates.

(ii) Mature trophozoite – small, amoeboid and intracellular.

(iii) Zygote non-motile, producing naked sporozoites.

(iv) Reproduction – alternately by schizogony in vertebrates and by sporogony in the invertebrate host.

Examples : (1) *Plasmodium* (causing malaria in man)

(2) *Haemoproteus* in flies, (3) *Babesia* in cattle,

(4) *Leucocytozoon* (causing disease in ducks).

Sub Class B. Neosporidia

1. Trophozoite – multinucleated ; spore cases generally containing a single germ.

Order 1. Myxosporidia

(i) They are found chiefly in cavities or tissues of fresh water fishes.

(ii) Trophozoite amoeboid but not intracellular.

(iii) Spores with 2 valves ; polar capsules 1 to 4.

Examples : (1) *Nosema bombycis* (in silk worms),

(2) *Nosema apis,* causing disease in honey bees,

(3) *Myxidium* (in fishes), etc.

Order 2. Sarcosporidia

(i) They are found generally in the striped muscles or tissues of mammals.

(ii) Body form-long, spindle like.

(iii) Spores – naked, no polar capsule, many in large cyst to 55 mm. in diameter.

Examples : (1) *Sarcocystis* (found in man and in muscles of pig),

(2) *Globidium.*

Class IV Ciliophora (Infusoria)

1. They are found in fresh or salt waters and in vegetable infusions.
2. Most of them are free-swimming and solitary and a few are parasitic, while others are sessile, and a few others form colonies.
3. They possess cilia throughout life or in some part of the life ; cilia serve for locomotion and food-getting. Cilia arise from the bottom or grooves or pits in the pellicle.
4. Each species is of constant and characteristic form and most of them have 2 nuclei – vegetative meganucleus, the other reproductive micronucleus.
5. Nutrition – holozoic.
6. Reproduction – asexual by binary fission, sexual by conjugation and autogamy.

Sub-Class (A) Ciliata :

1. Cilia present throughout life.
2. No tentacles.
3. Nuclei are of two types
4. Nutrition – holozoic.

Order 1. Holotricha

(i) Cilia simple and short, arranged in longitudinal rows, with gullet and undulating membrane . Adoral cilia absent or incospicuous.

(ii) Nutrition is holozoic or parasitic (in Astomata).

Examples : (1) *Paramecium*, (2) *Opalina*, (3) *Didinium*.

Order 2. Heterotricha

(i) With short ad-oral zone of cilia.

(ii) Contractile vacuoles when present, are fed by 2 long radiating canals.

Examples : (1) *Stentor*, (2) *Nyctotherus*, (3) *Balantidium*.

Order 3. Hypotricha

(i) Fresh living ciliates with a dorso ventrally flattened body ; dorsal surface is mostly devoid of cilia, and bears a few long stiff immobile bristles; larger and stiffer cilia are present on the lower surface of the body, which act in a way analogous to legs.

Examples : (1) *Euplotes*, (2) *Diophrys*, etc.

Order 4. Peritricha

(i) Mostly sessile ; body bell or vase like form with a stalk or peduncle.

(ii) Ciliation is reduced to the adoral zone in the disc like oral region.

Examples : (1) *Vorticella*, (2) *Epistylis*.

Sub-Class B. Suctoria

1. They are sedentary ciliates whose affinities are shown by the young stages.
2. They bear numerous rod-shaped protoplasmic "tentacles", adults are without cilia, sessile and stalked; cilia in only young stages.

(a) The tentacles spring from the whole surface or from the free end, or occur in groups, and have knob and pointed ends.

(b) The tentacles consist of an outer contractile sheath, which probably contains very fine myonemes and are thrown into folds on contraction, and an inner stiff tube, which extends deeply into the general endoplasm.

(c) The knob when present is permeated with fine radial canals. In some (e.g. Acineta), the knobbed tentacles are grouped into 2 or 3 clusters. When the tentacles come into contact with the prey, the swelling adheres to it, and the prey is soon killed.

3. The young suctoricans are budded off from the parent body (adult) and resembles ciliates in being ciliated and free swimming.
4. The cell body of the suctorians is spherical, elongate or branched in different species, usually attached by a stalk or disc to some objects, and covered by a pellicle or theca.
5. One or more contractile vacuoles are present on the body, but no cytostome.
6. Nutrition is holozoic, feeding on other Protista.
7. Generally with Mega and Micronuclei, the former exhibiting oval, elongated, monoliform (like a string of beads), or branched, whereas that of the latter (Micronucleus), may be spherical, sometimes oval or biconvex, usually consisting of central body encircled by a zone of nuclear sap.
8. Reproduction is by External (or Exogenous) or Internal (or Endogenous) budding with the production of a ciliated young called an embryo.Conjugation occurs in some species like other ciliates.
9. Mostly, they are parasitic, some inhabit fresh, brackish and salt waters.

Examples : (1) *Acineta*, (2) *Dendrosoma*,

(3) *Podophrya*.

Uncertain Class V of Protozoa :

Mycetozoa :

1. Mycetozoa are neither aquatic like other members of the Phylum Protozoa nor parasitic like many other forms, but they lead a sub-aerial life on decaying vegetable remains. Members of this group form encrusting masses in rotten wood and the like.
2. Body is naked with a large multinucleate plasmodium. Adult body is with many nuclei and numerous contractile vacuoles.
3. Nutrition – holozoic or holophytic.
4. Reproduction takes place by fission and by the formation of spores. The spores hatch out little amoebae which may acquire flagella. Dispersal may take place by the formation of multinucleate cysts.

Examples : (1) *Badhamia*, (2) *Didymium*,
(3) *Plasmodiophora*, etc.

Locomotion and Locomotory Organelles in Protozoa

Locomotion means change of place or movement or progression through the medium. The members of the Phylum Protozoa change their place and position, possibly in response to different life conditions. They exhibit a great diversity both in structure and habitat and consequently varied forms of locomotory organelles are found in Protozoa. The organelles, by which the change of place is effected are known as locomotory organelles. The seat of locomotion lies in the ectoplasm since the locomotory organelles either arise from it or are present in it.

The locomotory organelles in Protozoa include pseudopodia, flagella, cilia and pellicular contractile structures.

Modes of Locomotion in Protozoa

I. **Pseudopodial or Amoeboid Movement:** Some Protozoa move with the help of pseudopodia. The pseudopodia are blunt, finger-like temporary protrusions of the protoplasm of naked protozoa, which do not possess a definite pellicle, such as Rhizopoda (Sarcodina), many flagellates and a few sporozoa. Pseudopodia may be variously shaped. They may be formed of ectoplasm alone or they may also have a core of endoplasm. The actual formation of pseudopodium is brought about by a change in the colloidal form of the protoplasm at the point where the pseudopodium appears. The ectoplasm is in the gel condition and where the protuberance is to be formed, changes to the sol condition ; the general ectoplasm covering, owing to its elasticity contracts and forces the fluid endoplasm outwards. On reaching the surface, the endoplasm gels to reform the ectoplasm.

According to the "sol-gel" or "change of viscosity" theory of Hyman (1917) pseudopodial or amoeboid movement is effected by changes of colloidal protoplasm from a sol or fluid state to a gel or solid state, and vice-versa.

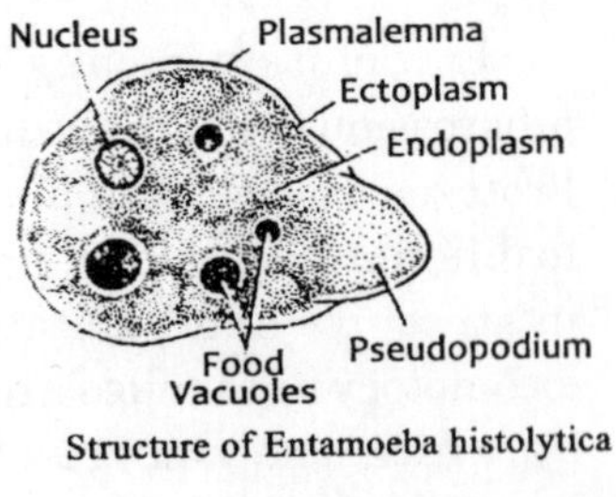

Structure of Entamoeba histolytica

According to Dellinger (1906), the newly formed pseudopodia attach their tips to the substratum and then contract their tips to the substratum and then contract, pulling the body forward. Dellinger described this movement in Amoeba as walking locomotion. According to Jennings (1904), amoeboid movement in *Amoeba verrucosa* is effected by a rolling movement of the body surface.

According to the surface tension theory of Berthold (1886), Verworns, Butschli and Rhumbler (1898), the pseudopodium is formed by an outflow of protoplasm from a weak point, where the surface tension becomes reduced by internal or external factors.

Types of Pseudopodia

According to the shape, size, structure and activity, 4 (four) kinds of pseudopodia are distinguished.

(i) **Lobopodia** : They are relatively broad, finger or tongue-like, sometimes branched, and their distal ends are typically rounded. They are made up of granular endoplasm and hyaline ectoplasm. They are quickly formed and equally quickly retracted. Ordinarily the formation of lobopodia is by a general flow of the cytoplasm, but in some it is sudden and 'eruptive' as in *Entamoeba histolytica*. Labopodia are common in Amoebae although they are also formed by certain flagellates. In *Amoeba proteus* several lobopodia may be given out from the body surface, while in others like *Amoeba limax,* the whole body flows into a lobopodium.

(ii) **Filopodia** : They are fine, slender, thread-like or filamentous, hyaline projections with pointed tips. They are composed almost exclusively of the ectoplasm. They may sometimes be branched, but the branches do not anastomose. They are commonly found in Testaceans, such as Euglyphya, Boderia, etc.

(iii) **Rhizopodia or Myxopodia or Reticulopodia** : They are filamentous, but branching and anastomosing. The branches of adjacent pseudopodia often form a network which serves for capturing prey. They are common in Foraminifera, such as Polystomella, Chlamydophrys, etc.

and in certain Testaceans, such as Myxotheca, Lieberkuhnia, etc.

(iv) **Axopodia**: They are stiff, spine-like and semitransparent pseudopodia. The axopodium is composed of an outer cytoplasmic envelope covering an axial rod or filament made of fibrils. Axopodia serve more for food capturing than locomotion. They are found in many Heliozoa, such as Actinophrys, Actinosphaerium, Compsonema, etc.

II. **Flagellar movement**: The flagella are characteristic of Flagellata (Mastigophora), but they are also found in flagellated stages of some Rhizopods and Sporozoa. Flagella are used for swimming, for creating food currents and for anchorage. They also act as sensory organelles for exploring the media.

Path of *Euglena* during flagellar movement.

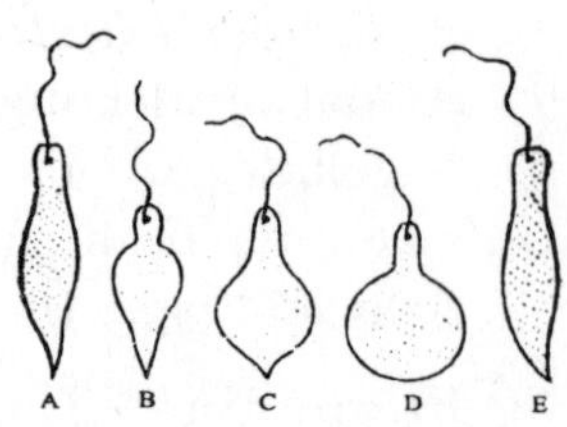

Stages in *Euglenoid* movment

A flagellum is a filamentous extension of the cytoplasm and is ordinarily very fine and highly vibratile. It is made up of two definite elements – an axial, highly vibratile filament (axoneme), which is formed as an outgrowth from the basal body or blepharoplast, and an enveloping contractile outer sheath which is formed from the periplast substance of the cortex. In some cases the sheath is circular in cross-section, in others ellipsoidal, while the contractile thread which is usually attached firmly to the sheath may run in a straight line the entire length of the sheath, or may follow a spiral course. In a few types, such as *Peranema trichophorum* or certain species of Bodo, the sheath remains passive, where its activity in the surrounding medium results in a steady progressive movement of the cell. In some flagella, the outer sheath may also contain fibrils or mastigonemes or "Flimmer". On the basis of the arrangement of the mastigonemes, the following types of flagella may be recognized :

(i) Stichonematic flagellum, having a single row of mastigonemes along one side of the sheath, as in Euglena.

(ii) Pantonematic flagellum, bearing two or more rows of mastigonemes, as in Peranema and Monas.

(iii) Acronematic flagellum, bearing a terminal filament or end piece, as in Polytoma, Chlamydomonas, etc.

(iv) Pentachromatic flagellum, bearing a terminal filament and one or more rows of mastigonemes.

(v) Simple flagella, having no terminal filament or mastigonemes, as in Dinoflagellata.

On the basis of the behaviour, the flagella may also be of two types:

(i) **Tractellum**: The flagellum is inserted at the anterior end of the body and is directed anteriorly. When the flagellum moves, the body is pulled forward.

The posteriorly directed flagellum is known as trailing flagellum.

(ii) **Pulselium**: The flagellum is inserted near the posterior end of the body. Its vibration pushes the animal forward.

Flagellar movement

1. According to Butschli's *Screw Theory,* the flagellum undergoes a series of lateral movements driving the

animal forward, followed by a spiral rotation at its own axis.

2. According to Verworn the flagellum moves in a wave-like undulating motion, as in *Peranema trichophorum*. The animal shows an extremely rapid forward movement when the flagellum is thrown into sinuous undulation proceeding from tip to base. However, the animal moves forward slowly if only the distal part of the flagellum shows undulations. If, on the other hand, undulations progress from base to tip, the animal shows a backward movement. Likewise, spiral undulations cause rotation in the opposite direction.
3. Delage and Herouard have stated that a forward pull results from the combination of circular movements and the cell either glides smoothly after its active propeller or rotates more or less rapidly on its long axis while freely swimming. If two flagella are present a curious shaking movement may accompany rotation and translocation.

Whatever may be the mode of flagellar movement, the flagellates travel swiftly through water. The speed of flagellar movement may range from 15 to 300 micra per second.

III. Ciliary Movement

Ciliated Protozoa move with the help of cilia. Cilia are characteristic of ciliata and larval suctoria. Among the Protozoans, the Ciliates are the fastest, moving at the rate of 400 to 2000 micron per second.

Cilia are small, slender and more or less short hair-like processes of ectoplasm of the body. They are different from flagella in that they are much shorter compared to the size of the body. Their number is very large, for instance, Paramecium possesses 10,000 to 14,000 cilia all over its body, while *Porodon teres* has about 11,600 cilia on its body.

A cilium consists of an axial filament or axoneme surrounded by an elastic sheath continuous with the plasmamembrane of the cell surface. The axoneme consists of a paired fibres and 2 central fibres. The central fibres are enclosed within a delicate inner sheath.

Cilium arises from a basal granule or blepharoplast situated in the ectoplasm. According to Henneguy (1898), the basal granules are centrioles or their derivatives.

The arrangement of cilia on the surface of the body differs from species to species. They may form a complete coating for the organism as in many Holotrichida, or they are limited to certain zones as in *Urocentrum turbo, Didinium nasutum* (Fig.) or to the ventral surface as in generalised Hypotrichida. In Vorticella, cilia are restricted only in a concentric circlet at the distal end of the body. In some cases, for instance, Opalina, cilia are uniform in length over the entire body ; in other cases they are longer in the region of the mouth. In all cases cilia are arranged in longer or shorter rows running straight or spirally, and giving the stripped appearance. Waves of contraction pass from the anterior end posteriorly, cilia of the same transverse rows beating synchronously, those of the same longitudinal rows metachronously.

Combinations of cilia may give rise to membranulae, membranelles, undulating membranes and cirri, each of which originates as an independent organoid.

(i) **Membranulae**: They are very long, delicate, finely pointed aggregates of cilia which differ from the somewhat similar cirri in movements and in composition, while their basal granules in *Didinium nasutom* at least, are connected with the vicinity of the nucleus by definite rhizoplasts.

(ii) **Membranelles**: They are formed by the fusion of cilia in the mouth region, and their flat curved faces make powerful sweeps in the water so as to bring the food particles in the mouth (cytostome) as well as to help locomotion of the animal. In the Vorticellidae there are two rows of membranelles.

(iii) **Undulating membranes**: They are formed of one or more longitudinal rows of cilia which have fused together. In some simplest cases these membranes are composed of a single row, longitudinally placed cilia. In some complex forms undulating membranes may be composed of 3 to 10 rows of cilia fused in longitudinal rows.

Undulating membranes are found in all orders of Ciliates, and they range in size from delicate aggregates no broader from base to tip then ordinary cilia to relatively enormous balloon-like structures equal in width to more than 2 the diameter of the body. In some, as in *Lembadion conchoides*, this membrane is almost equal to the length of the body.

(iv) **Cirri**: Cirri are the most highly specialised of all the motile organs of Ciliates. A cirrus is a large, stiff, bristle-like organ composed of a tuft of cilia arranged in two or three rows, fused completely. The cirri are termed frontal, anal, caudal, ventral, marginal, etc., according to their positions in the body. In the cross-section near the base, cirri are either circular, ellipsoidal, quadrilateral or irregular. They have a basal plate made up of the basal granules of the fused cilia.

The cirri are movable in all directions. Their stroke is not confined to one plane but may be in any direction. This gives to the Hypotrichida an extreme variety of movements. Many of them walk on the tips of their frontal and ventral cirri (Stylonchia); others swim with a peculiar jurky movement (Aspidisca) ; still some others combine swimming with sudden jumps or springs due to anal or caudal cirri (Uronychia, Euplotes).

In addition to the above locomotory organelles, some Protozoa possess Myonemes or Myofibrils. Primarily these are for metaboly and secondarily for locomotion. Myonemes are very fine and highly contractile fibrils, running in various directions in the ectoplasm of the body. They may be in the form of ridges and grooves (e.g., Euglena), micro-tubules (e.g. Trypanosoma), miofibrils (e.g. Ciliates) or band-like and cross-striated (e.g. Stentor). The longitudinal myonemes of the body may converge basally into the stalks forming the spasmonemes (e.g. Vorticella). Movement effected by myonomes is known as Peristaltic movement.

Ciliary Movement

During locomotion, cilia bend like a pendulum throughout their length to strike the water which moves in the direction of the beat while the ciliate tends to move in opposite direction. All the body cilia do not pulsate simultaneously. A wave of pulsation passes from one end of the body to the other, and in this, cilia of a longitudinal row beat metachronously, while cilia of a transverse row beat synchronously. In most cases, ciliates move in a spiral path, rotating on their own axis (e.g. Paramecium).

Illustration of metachronous movement in a single row of cilia

Two stges in ciliry movement :
A - Effective stroke.
B - Recovery stroke.

The mechanism of ciliary movements and the movement effected by the derivation of cilia is considered to be controlled by a conductile neuromotor system made up of a central motorium and the basal bodies connecting the cilia.

OSMOREGULATION IN PROTOZOA

Osmoregulation is a process in which water contents of

the protoplasm are regulated and balanced at a proper level. The contractile vacuoles present in the bodies of fresh water Protozoa and some marine and parasitic ciliates are concerned with the osmoregulation.

In Amoeba, the body protoplasm is more concentrated than the surrounding water, and so, water enters regularly in its body by osmosis through the semipermeable plasmalemma. Water may also enter the animal body as a result of respiration, or water may also be included when food particles are engulfed. If the excess water is not eliminated, it would cause rupture of its body. The excess amount of water is then removed by the contractile vacuole.

First, the contractile vacuole appears as a small circular structure in the cytoplasm. It gradually increases in size by the collection of more liquid in it. When fully formed, a contractile vacuole assumes as a large, round, clear and pulsatile sac (e.g. Amoeba). After attaining a maximum size, the contractile vacuole contracts and expands, and then collapses to invisibility, discharging its contents to the exterior through a temporary pore on the surface. The intervals between successive contractions vary from a few seconds to several minutes in fresh water Protozoa. The rate of contraction is subject to change with the physiological state of the organism, amount of food, temperature, osmotic pressure of the medium, etc.

According to Rossbach, the contraction of the contractile vacuole is accelerated either rapidly or more slowly with the rise of the temperature in the case of the following 3 (three) ciliates mentioned below :

	Time in seconds between two systoles at different temperature (C)					
	5^0	10^0	15^0	20^0	25^0	30^0
Euplotes charon	61	48	31	28	22	23
Stylonchia pustulata	18	14	10-11	6-8	5-6	4
Chilodonella cucullulus	9	7	5	4	4	–

As soon as the old contractile vacuole disappears, new one appears again from one or more minute droplets. Gradually it grows in size as before (diastole), until it bursts through the thin cytoplasmic layer above it (Systole), discharging its contents to the exterior.

In the case of marine Amoeba, contractile vacuole is absent, but if it is kept in fresh water, it develops the contractile vacuole. The contractile vacuole is altogether absent in Sporozoa.

In the Flagellata, the contractile vacuole appears to be located in the anterior region of the body, while in Ciliata it is located in the deepest part of the ectoplasm, and so, constant in position. However, in other forms of Protozoa, the location of the contractile vacuole is not definite. It moves about with the cytoplasmic movements.

The number of contractile vacuole in the majority of Protozoa may be one to many. In *Paramecium multinucleatum,* there are 2 to 7 contractile vacuoles located on the aboral side below the ectoplasm. Each vacuole has a permanent pore, and leading to the pore is a short tube-like invagination of the pellicle. When only one contractile vacuole is present, it is usually located either near the cytopharynx or, in the posterior part of the body. When many vacuoles are present, they are distributed in linear series in the body. In *Paramecium woodruff,* the contractile vacuoles are deeply seated, and a delicate duct connects the vacuole with the pore on the pellicle. In Nyctotherus, Balantidium, etc., the contractile vacuole appears very close to the permanent cytopyge located at the posterior extremity, and its contents are discharged through it.

In many Ciliates, radiating or collecting canals are present in addition to the main contractile vacuole. The collecting canals may radiate from the central canal as in Paramecium, Frontoria, etc., but when the central vacuole is terminal, the canals do not radiate. The number of collecting canals may be one to many. In *Paramecium multinucleatum,* each vacuole has 5-10 long collecting canals, made up of terminal portion, a proximal injection canal and an ampulla between them. The injection canal is connected with the pore and the fluid is expelled through the pore.

NUTRITION IN PROTOZOA

Nutrition is a physiological process involving ingestion (taking in of food), secretion and digestion, absorption, assimilation and defaecation. Protozoan animals are at the root of the animal kingdom, and they bridge the gulf between the plant and animal world in one hand and also the unicellular and multicellular organisms on the other, which is more or less evidenced by their various modes of nutrition, such as holozoic, holophytic, saprophytic, parasitic, mixotrophic, etc.

1. **Holozoic (zootrophic, heterotrophic) nutrition**: This is in other words true animal-like nutrition involving all physiological processes of an well organized digestive system but in a similar form and in detail just an opposite to the holophytic type of nutrition. Thus this holozoic type of nutrition in Protozoa is like that of Metazoa, and it involves the ingestion of raw materials in the form of proteins, carbohydrates and fats which are usually combined in the protoplasm of some other living organism eaten as food.

   Rhumbler (1910) distinguishes 4 methods of food ingestion in Rhizopoda: (a) Circumvallation, (b) Circumfluence, (c) Invagination, and (d) Importation.

   (a) *Circumvallation*: Here the prey (food) is captured by throwing out a wide food-cap that embraces the prey without touching it, e.g. Amoeba.

(b) *Circumfluence*: It appears to be due to a stimulus emanating from a living body, the effect of which (Jennings, 1904) is to cause pseudopodia to flow toward the prey and to entrap it as in Axopods and Ruticulopods, etc.

(c) *Invagination*: Here the food particle is drawn into the deepening depression of the cytoplasm and ultimately the food is sucked into the endoplasm, e.g. *Amoeba terricola*.

(d) *Importation*: Here the food is drawn into the protoplasmic body as soon as it comes in contact with it, e.g. *Amoeba dofleini*.

Food Channel

In Flagellates and Ciliates sub-serving holozoic mode of nutrition, a definite region becomes specialised for food intake. In its simplest form this consists of a region at the base of the flagellum, e.g., some flagellates. In the next stage there is a simple mouth opening or cyclostome, primitively terminal but often shifted to lateral position. In raptorial forms, the food passage is supported by a circlet of trichites termed the pharyngeal basket (e.g., *Nasula aurea*). Then a tubular passage, the cytopharynx, also called gullet and oesophagus develops extending from the surface deep into the endoplasm. Lastly, a portion of the body surface, the peristome, may be specialised for directing food into the cytopharynx, and both cytopharynx and peristome come to be provided with special ciliary tracts or undulating membranes for creating water current that sweep the food into the cytophayrnx. In some forms e.g., *Pycnothrix monocystoides,* a definite anal apparatus is developed. Sharp describes this as rectum with distinct walls opening to the outside by a permanent cytopyge, while at the inner end there is a 'cecum' which acts as a collecting vacuole for the faecal matter.

In Paradox, mouth remains closed, opens only when food is engulfed. In Paramecium, mouth is permanently open with a well marked peristome. Anus is temporary. In Vorticella, mouth, peristome and vestibule are present. Anus is temporary. In Nyctotherus, both mouth and anus are distinct openings and permanent.

In Suctoria, there is no mouth but they are provided with tentacles which secrete a toxic substance to paralyze the prey.

Many Protozoa exhibit discrimination in food selection. For instance, *Didinium nasutum* has a special liking for Paramecium. *Actinobolus radians* selects only *Halteria grandinella.*

Mode of feeding in Protozoa

The majority of the Protozoa are holozoic in their methods of food-getting, and we can distinguish two main groups: (i) the continuous feeders, and (ii) the occasional feeders.

(i) The continuous feeders are those forms with permanently open mouth through which a constant stream of water current is maintained by the action of the peristomial motile apparatus.

(ii) The occasional feeders like carnivorous types of Metazoa, feed whenever chance brings prey within the radius of their activity. In some cases, babloon-like membranes are unfolded and spread out like sails for the direction of food current to the mouth, e.g., Pleuronema chryalis. In other cases great net-like traps are spread for the capture of unwary diatoms, etc. In some again, e.g., Actinobolia radians, when at rest protrudes a forest of radiating tentacles the end of which carry trichocysts and thus help in capturing prey.

Digestion and Assimilation

In many Protozoan species, the food organisms are paralyzed or killed upon contact with pseudopodia, tentacles or exploded trichocysts. Digestion is mostly intercellular in nature. The ingested food materials are usually surrounded by a food vacuole which contains water and some secretion of the endoplasm. The food vacuoles reach the endoplasm and they move along a definite course by endoplasmic currents. In Ciliates (e.g., *Paramecium*), the movement of the food vacuoles is along the aboral side to the anterior region and down the other side, with a short cyclosis in the posterior half of the body. In some other Ciliates, there is a definite "digestive tubule" beginning with the cytostome and ending in the cytopyge, and the food vacuoles travel through it. In *Vorticella*, food vacuoles are formed one by one at the end of cytopharynx and then migrate through different parts of the cytoplasm without order.

During digestion of food, the food vacuoles pass through an acid phase, followed by the alkaline phase. In the acid phase, the prey is killed and the contents of the vacuole are concentrated. Digestion occurs mostly during the alkaline phase. Several digestive enzymes are provided by the lysosomes, and almost all the food stuffs are said to be digested by them.

The digested food is absorbed and assimilated and supplied to every part of the cell body by means of diffusion. The excess food may be stored as lipids, glycogen, paramylon, chromatid bodies, etc.

The indigestible residue of the food is egested or defaecated from the body. In Rhizopodia, egestion of the residue takes place at any point. It is emitted and left behind as the animal flows onwards. In pellicle bearing forms, the egestion takes place either through the cytopyge located at the posterior region of the body or through an aperture to the vestibule.

2. **Holophytic or Autotrophic or Phytotrophic nutrition**: This type of nutrition is identical with photosynthesis of plants in which carbohydrates are synthesised from water and carbon-dioxide with the aid

of chlorophyll and sunlight. The products of holophytic nutrition are amyloid substances which are formed in the chromatophores around certain small bodies known as pyrenoids. Holophytic nutrition is widespread among the chlorophyll bearing flagellates (e.g. Euglena).

3. **Saprozoic or Saprophytic nutrition**: In this type, the Protozoa subsists in a medium of entirely dissolved substances. The dissolved substances are absorbed through the entire surface or possibly, through limited regions of the body by 'endosmosis'. The dissolved food matters include proteins and carbohydrates produced by the disintegration of animals and plants by bacteria. Saprozoic nutrition is found in the majority of parasitic Protozoa and in a number of free living colourless flagellates, e.g. *Trypanosoma* and *Mycetozoa*.

4. **Parasitic nutrition**: This type of nutrition is effected by the absorption of nutrient materials from other organisms called the host. Sporozoans are entirely parasitic. There are two ways by which parasitic Protozoa obtain food: (a) feeding upon the raw material or digested food of the host. (b) feeding upon the living substance of the host body.

5. **Coprozoic nutrition**: Some free living Protozoans such as Rhynchomonas, Chlamydophyrs, etc., feed upon the faecal matters of other organisms. This mode of nutrition is termed Coprozoic.

6. **Myxotrophic nutrition**: Many Protozoa nourish themselves by more than one method at the same or different times, depending upon the environmental conditions. This type of mixed nutrition is referred to mixotrophic nutrition. *Euglena gracilis*, for instance, can nourish itself by holophytic mode in day and by saprophytic mode in the darkness.

REPRODUCTION IN PROTOZOA

Reproduction is a process by which new individuals like the parents are produced from the pre-existing organisms. This is a unique property possessed by the living organisms alone. Its manifestations are many and varied from the simple division of the parental body into two or more parts to the extremely complicated reproductive processes of higher animals.

Protozoa reproduces both by asexual and sexual means.

I. **Asexual Reproduction**: In this mode, there is no union of gametes; only one animal can produce new individuals. There are 4 types of asexual reproduction in Protozoa : binary fission, multiple fission, plasmotomy and budding.

1. **Binary fission**: This is a mode of division of the body into approximately equal parts. In the process, the nucleus divides into two by a simplified form of mitosis, then the cytoplasm divides into two equal halves by constriction. The wanting structures which cannot be equally shared by both the offsprings can again be reformed by both offsprings.

   Binary fission may be of the following types : transverse binary fission, longitudinal binary fission, oblique binary fission.

   (a) Transverse binary fission: In this the fission plane is transverse, as in *Amoeba*, *Paramecium*.

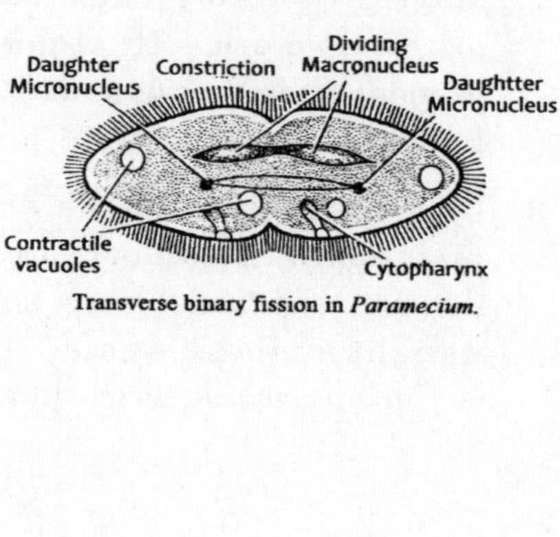

Transverse binary fission in *Paramecium*.

   (b) Longitudinal binary fission:

In this the fission is longitudinal starting from the anterior end, e.g. *Euglena*, *Vorticella*, etc.

(c) Oblique binary fission: In this, the fission plane is somewhat oblique, e.g., *Ceratium*.

2. Multiple fission or Sporulation: In this many individuals are produced by one at a time, e.g., *Amoeba*, *Euglena*, Polystomella. In the process, the nucleus first divides repeatedly and a large number of daughter nuclei are produced. These nuclei are usually arranged themselves at the periphery. Then the cytoplasm breaks up, and a small bit of it surrounds each daughter nucleus, forming minute animals. Under favourable conditions, these small animals come out and grow to adult stage.

   Multiple fission is termed *Schizogony* or *Agamogny* when occurring in an asexual cycle (e.g., *Plasmodium*), producing agametes or merozoites each of which directly develops into an adult individual.

   Multiple fission is called Sporogony when following sexual fusion, and their products are called spores when enveloped by a cyst wall.

   Multiple fission is termed Gamogony when it occurs in a sexual cycle and the products are gametes or sex cells (e.g., *Monocystis*).

   Motile products of multiple fission are called Swarm spores or Swarmers; if amoeboid they are called pseudopodiospores or amoebospores; if flagellate, they are known as flagellospores.

3. Plasmotomy: It is the division of a multinucleate protozoan into two or more parts without any nuclear division, the nuclei being simply distributed among the daughter individuals. This occurs in *Opalina*, *Actinosphaerium*, *Pelomyxa*, etc.

4. Budding or Gemmation : In this type, an outgrowth takes place on the body of the animal. The outgrowth (bud) receives a portion of the parental nucleus, dissociates from the parent body and develops into the normal adult.

Budding may be simple budding when a single bud is formed at a time (e.g., *Vorticella*); multiple budding, when several buds are formed simultaneously (e.g. *Suctoria*); exogenous budding, when one to many buds are constricted off from the outer surface of the parent body, as in *Ephelota*, *Noctiluca*, etc.; endogenous budding, when buds are liberated inside the parent body into internal spaces or brood chambers, and later they come out to the exterior by developing cilia.

II. Sexual Reproduction

The mode of reproduction in which two kinds of gametes unite to form a new individual is known as sexual reproduction. In Metazoan animals, sexual reproduction involves liberation of special reproductive cells from the body and fertilization of ovum by spermatozoan.

In Protozoa, two types of sexual reproduction occur : (1) capulation or syngamy, involving the fusion of two sex cells; and (2) conjugation or temporary union of two gametes or individuals of the same species followed by interchange of nuclear materials.

1. **Syngamy**: In this, there is complete and permanent union of two protozoan animals, or gametes, resulting in the formation of zygote. Depending upon the degrees of differentiation of the gametes, syngamy may be of the following types :

(a) Autogamy: When the fusing gametes are products of one parent cell, the union is known as autogamy. It occurs in Actinophrys and Actinosphaerium.

(b) Hologamy: Sexual fusion of two ordinary mature

protozoan individuals behaving themselves as gametes, is termed hologamy. It occurs in a few Sarcodina and Flagellata.

(c) Paedogamy: When the fusing individuals are young, the phenomenon is known as paedogamy. It occurs in Actinophrys.

(d) Exogamy: When fusion (Syngamy) takes place between the offsprings of different parents, the phenomenon is termed exogamy.

(e) Merogamy: In this, the uniting gametes or microgametes are small and they are formed by a special method of fission.

(f) Isogamy: When the copulating gametes are similar in size and form, they are called isogametes and their union is isogamy. Isogamous gametes are generally produced by multiple fission; they often escape as flagellate swarmers and fuse in pairs. This occurs in Foraminifera (e.g. *Polystomella*), and Gregarines (e.g. *Monocystis*).

(g) Anisogamy: When two fusing gametes differ in size, morphology and behaviour, they are termed anisogametes, and their fusion is known as anisogamy or heterogamy. This is commonly found in Sporozoa (e.g., *Plasmodium*, *Eimeria*), Phytomonadina (e.g. *Volvox*).

(h) Merogamy: This is syngamy taking place between large gametes (macrogametes) of the species. It is seen in *Arcella*.

(i) Microgamy: It is the union of two smaller gametes as is found in *Arcella*.

The result of the fusion of two gametes is the formation of the zygote and its nucleus formed by the union of the two nuclei of the gametes is called the synkarion. The zygote may develop directly into the adult or, may undergo encystation and then emerge as a young animal. It may also undergo multiple fission within the cyst, and finally they escape from the cyst wall and grow to the adult structure.

2. Conjugation

Conjugation is a temporary union of two adult individuals of the same species. This is commonly found in Ciliata (e.g., *Paramecium*, *Vorticella*) and Suctoria. The associating sexual individuals are designated as conjugants which may be equal (isogamous) as in *Paramecium* or unequal (anisogamous), as in Vorticella. In the process, the conjugants interchange their nuclear materials; the stationary or female pronucleus in each conjugant fuses with the migratory or male pronucleus coming from the other conjugant, and thus a zygote nucleus is formed. Then the two conjugants separate as ex-conjugants. Each of them again undergoes nuclear and cytoplasmic divisions, producing 4 daughter individuals with the original nuclear condition as in *Paramecium*.

Factors of conjugation

Various environmental factors, such as temperature, light, food, bacterial strength, culture medium, etc., affect conjugation. According to Maupas, individuals undergo a certain number of asexual reproduction before conjugation takes place. Conjugation could, as well, be induced by starvation following rich feeding.

Significance of conjugation

Protozoans are also subject to periodical waxing and waning of vitality (Senescence). If reproduction by binary fission is continued for several generations, the individual tends to decay in the purity of the race; the individual loses its physiological efficiency, becomes reduced in size, diminishes its vigour, retards its organization and enters into a state of depression. And if this state of thing is continued for long, ultimately death semms 'inevitable'. To avoid this, conjugation is resorted to, and the nuclear materials between the two individuals are exchanged so that there tends to rejuvenate the old organisms. Conjugation is also a process of reorganization leading to the released vigour of all vital activities including reproduction.

3. **Automixis**: This is fusion of two gametic nuclei derived from a single nucleus of an individual. Automixis may again be of – (a) Autogamy, when two gametic nuclei (pronuclei) which undergo fusion, are present inside the same cell (as in Myxosoma); (b) Paedogamy, when two pronuclei present in two different cells derived from a single parent cell fuse (as in *Actinosphaerium eicchorni*); (c) Cytogamy, when conjugating pairs are formed without interchanging pronuclei (as in *Paramecium caudatum*).
4. **Endomixis**: It is a phenomenon of replacement of the macronucleus (meganucleus) and the nuclear reorganization in a normal individual (without nuclear contribution from another individual). It occurs in Paramecium caudatum. It is said to occur periodically and monthly, when conjugation is prevented, that is, when two individuals are checked from meeting. Endomixis resembles conjugation in the replacement of the macronucleus from micronuclear material. But endomixis also differs from conjugation in not occurring union of the nuclei, thereby not having recombination of hereditary materials.
5. **Plasmotomy**: This is a non-sexual process. Here two or more individuals may fuse by their cytoplasm to form a plasmodium; later they separate again unchanged.
6. **Parthenogenesis**: Here the gametes which fail at cross-fertilization grow and develop parthenogenetically, as in *Chlamydomonas*.

7. **Regeneration**: Many Protozoans possess the capacity to replace or regenerate the lost parts. Free-living Rhizopods having nucleated piece can reform the missing parts thereby assuming the normal shape, often within a few hours.

STRUCTURES AND LIFE HISTORY OF AMOEBA

(Phylum Protozoa, Class Rhizopodia, Order Lobosa)

The genus Amoeba has 6 important species : *Amoeba proteus, A. verrucosa, A. discoides, A. radiosa, A. dubia* and *A. terricola.* They live either in marine or freshwater.

***Amoeba proteus* – Structure**

It is found in fresh water, ponds, ditches, etc. Amoeba is a microscopic, one-called or unicellular animal. It is only about $\frac{1}{100}$ of an inch or 200 or 300 microns in diameter. When seen under the microscope, Amoeba appears as an irregular translucent mass of jelly-like substance. It undergoes frequent change of form by protruding and withdrawing little finger-like processes called pseudopodia or false feet.

The body consists of

(i) **Plasmalemma**: It is extremely delicate and elastic membrane making the boundary of the body. It retains the protoplasm within the cell but permits passage of water and certain gases.

(ii) **Ectoplasm**: This is a non-granular and clear transparent layer, lying below the plasmalemma. It lends form to the cell body.

(iii) **Endoplasm**: It lies next to the ectoplasm and enclosed by it. It is granular and transparent layer, more fluid than ectoplasm and forming the central parts of the cell body. Endoplasm has an outer stiff plasmagel, and an inner fluid plasmasol. Endoplasm as a whole assits in locomotion of the animal and maintains many other structures within it. The structures are :

(iv) **Nucleus**: Usually it lies in the centre of the endoplasm, although it has got no fixed position in the body. It is more or less disc-shaped, surrounded by a thin nuclear membrane inside which are masses of chromatin granules, suspended in the fluid portion of the nucleus, called nuclear sap. Nucleus controls the life of the organism; it brings about the growth, manufacture and use of food stuffs and formation of new Amoeba.

(v) **Contractile vacuole**: This is a clear bubble-like structure, lying near about the nucleus in the endoplasm. It is filled with watery fluid and bubbles of gas. At regular intervals, this vacuole moves to the surface, contracts and discharges its contents to the exterior and then reforms. The contractile vacuole controls the water content of the protoplasm (osmoregulation); it also serves as a respiratory and excretory organ.

(vi) **Water vacuoles**: These are spherical-shaped structures scattered in the endoplasm. They look like transparent drops of water. They do not change their shape. They also assist in water regulation of the organism.

(vii) **Food vacuoles**: These look like minute drops of water containing water and food in various phases of digestion. Food particles taken by the animal are stored and digested in the food vacuoles.

(viii) A few sand particles and small droplets of fat may also be present in the endoplasm. They are probably concerned with the reservation of fats, etc.

Physiology

1. **Locomotion**: Amoeba moves by forming and extending pseudopodia. The formation of pseudopodia is effected

by a change in the colloidal form of protoplasm from gel to sol condition and vice versa ('Change of viscosity' theory). First of all, towards certain point of the body, there appears a blunt projection of ectoplasm. The plasmagel layer lying at the root of the ectoplasmic projection becomes liquefied and is slowly transformed into plasmasol. The central portion of the cytoplasm, that is, plasmasol streams in the ectoplasmic projection. Now the stiff plasmagel lying at the hinder end is changed into fluid plasmasol and rushes at the forward end into the projection. This plasmasol is converted into ectoplasm. A repeatation of this process results in the formation of pseudopodia. The pseudopodia grow, and the animal moves. This mode of movement is called "Amoeboid movement".

2. **Nutrition**: The mode of nutrition in Amoeba is holozoic. It feeds on microscopic organisms, such as algae, other protozoans, rotifers and nematodes. It exercises choice of food to a certain extent.

Depending upon the nature of the food, ingestion in Amoeba may take place by the following methods :

(a) **Circumvalation**: When Amoeba comes in contact with the food material it wants, it sends out pseudopodia that encircle the food, which with some water, is taken into the endoplasm as a food vacuole.

(b) **Circumfluence**: When Amoeba comes in contact with an inactive prey, the pseudopodium forms a cup above the food and press it down to the substratum ; the cups encloses the food to form a food vacuole.

(c) **Invagination**: In this the prey is paralysed or killed by the toxic and sticky secretion of the pseudopodium. The ectoplasm along with the food is invaginated into the endoplasm in the form of a tube. The food is sucked into the tube.

(d) **Import**: Here the food sinks passively into the body.

The whole process of food-taking requires 1-2 minutes, depending on the nature of the food as well as temperature.

During digestion, the food vacuole moves from one part to another of the cytoplasm. Digestive fluids (enzymes) flow into these vacuoles. The fluid in the food vacuole is at first acidic and later alkaline. Proteins, fats and carbohydrates are broken down into simple soluble forms. The digested food is absorbed into the surrounding cytoplasm. The absorbed food serves for growth, movement and reproduction of the animal. The vacuole diminishes its size until only undigested matter remains. The undigested matter is egested from any point on the Amoeba's surface when it moves away from the particular place.

Excretion, Osmoregulation: Excretion in Amoeba is effected by diffusion through the general surface of the body. Sometimes contractile vacuoles also take part in the excretory activities. Small amount of the nitrogenous waste matters and the excess of water formed during the course of metabolism, gradually accumulates in the contractile vacuole. When the contractile vacuole is fully formed in this manner, it suddenly bursts, discharging its contents to the exterior. In this way, the contractile vacuole also performs osmoregulatory function.

Respiration: In respiration, free interchange of oxygen dissolved in water and carbon dioxide formed in the body is carried out by diffusion at the surface of the animal.

Response to stimuli: Amoeba responds to many stimuli, such as due to change in contact, light, etc. When Amoeba comes in contact with a food item, it responds positively. But if a creeping Amoeba is slightly touched with a needle, it shows negative response. Amoeba also reacts negatively to almost all chemicals and strong light.

Reproduction: Amoeba reproduces by simple asexual method which includes (i) Binary fission and (ii) Multiple fission or Sporulation.

(i) **Binary fission**: When Amoeba grows and attains a certain size, its nucleus first divides into two. Then the cytoplasm constricts across the middle line between the nuclei ; and the constriction deepens inward dividing the cytoplasm into two halves. Each half flows round the respective daughter nucleus to form a daughter Amoeba. Each daughter Amoeba reconstructs parts wanting in it and grows into a mature individual within three days. After this, it will be ready for the next binary fission.

(ii) **Multiple fission**: This takes place under unfavourable conditions. Under such conditions, Amoeba withdraws its pseudopodia and secretes a cyst wall around it. Within the cyst, the nucleus divides repeatedly to produce a large number of small nuclei and arrange themselves round the periphery. The cytoplasm breaks up into as many pieces as there are nuclei. Each small nucleus becomes surrounded by a bit of cytoplasm, forming a small cell, called pseudopodiospores. Under favourable condition, the cyst wall is broken and the pseudopodiospores are released. These swim with the help of little pseudopodia, feed and grow to become the adult.

Plastogamy: Sometimes many Amoeba fuse to form a single mass in which the cytoplasm from all the cells unite together but the nuclei remain discrete and they do not fuse. This resulting multinucleated mass is called plasmodium, and the process of its formation is called plastogamy. After sometime the plasmodium separates into component parts and each leads an independent existence.

ENTAMOEBA : STRUCTURE, LIFE HISTORY AND PATHOGENICITY

(Phylum Protozoa, Class Rhizopodia, Order Lobosa)

The genus Entamoeba has four important species: (1) *Entamoeba histolytica*, causing Amoebic dysentery in man, (2) *Entamoeba hartmani*, causing Amoebic dysentery, but is less harmful than *Entamoeba histolytica*, (3) *Entamoeba coli*, causing gastrointestinal troubles in man, (4) *Entamoeba gingivalis* causing abscesses of gums, throat and tonsils in man.

Entamoeba histolytica: Structure, life history and pathogenicity.

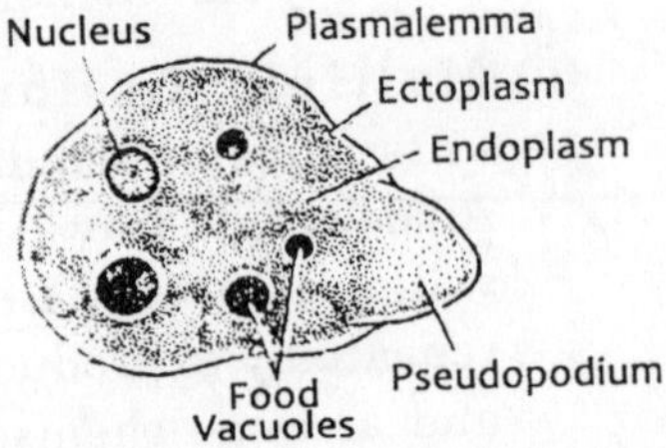

Structure of Entamoeba histolytica

Entamoeba histolytica is a microscopic parasitic protozoan, first described by Lambl (1859). Its pathogenic nature was discovered by Losch (1875). It is cosmopolitan in distribution. It attacks the mucosa and sub-mucosa of the intestinal wall of the human body, and causes amoebiasis or amoebic dysentery.

Entamoeba histolytica is unicellular. In its life cycle, it passes into three stages or forms, viz., (a) trophozoite, (b) precystic and (c) cystic.

(a) Trophozoite: It is the feeding and growing stage. It looks like a transparent and irregular mass of living substance, about 20-30 m in diameter. The cytoplasm has an outer hyaline ectoplasm and an inner granular endoplasm. Endoplasm forms a single blunt pseudopodium in the direction of movement. Endoplasm contains a round nucleus with a central endosome, and food vacuole in the form of bacteria or blood corpuscles. There is no contractile vacuole in the body.

(b) Precystic stage: It is non-feeding, spherical form having a blunt pseudopodium and a nucleus. Its size varies from 10 to 20 m in diameter. It lives in the lumen of the large intestine. It undergoes encystation.

(c) Cystic stage: The precystic form becomes rounded during encystation and is surrounded by a cyst wall. The mature cyst is a spherical body, and its cytoplasm contains reserve food and chromatoid bodies.

The nucleus of the cyst becomes divided into binucleate, tetra-nucleate or quadrinucleate. It then passes on with the faeces of the patient and forms the infective stage.

Life cycle

Reproduction: *Entamoeba histolytica* takes place in the trophozoite stage. It reproduces by binary fission within the wall of the large intestine. Under certain conditions, it also encysts. For this, the parasite becomes smaller and rounded. Its nucleus divides twice. An encysted tetranucleated Entamoeba is formed. It then forms a resting stage but it is highly infective.

The encysted parasite passes on with the faeces of the host. It may become contaminated with vegetables, water, etc., and if swallowed by another man, it gets ruptured in the intestine and released as young individuals (amoėbula). They grow at the expense of the host. They damage the colon tissues and cause ulcers discharging blood and mucus when the faeces become thin. If the growth of the parasite is not checked early, it may go to the liver, lungs, spleen and brain damaging them.

***Entamoeba histolytica:* Pathogenicity**

Parasitic organism like the *Entamoeba histolytica* causes

some disease to its host. The effects or symptoms of the disease are called pathological effects, and the phenomenon is termed pathogenicity.

Entamoeba histolytica causes dysentery, diarrhoea, hepatitis and liver abscesses in man. The term "amoebiasis" is applied to all these conditions caused due to infection of *Entamoeba histolytica*.

(i) Amoebic dysentery: In this the infection is confined to the intestine. The parasite feeds on the tissue cells causing necrosis of the epithelial lining of the colon so as to form abscesses which later burst to become bleeding ulcers. Under acute condition, the wall of the colon may become perforated, liberating huge number of bacteria into the abdominal cavity, leading to death. The patient suffering from amoebic dysentery has repeated, blood mixed, slimy and foul smelling motions.

(ii) Chronic intestinal amoebiasis: Persons with strong body resistance, or having suffered with repeated attacks of amoebic dysentery develops a chronic course of the disease. Such a person suffers from irregularity of bowel, abdominal pain, headache, nausea, loss of appetite, fatigue, nervousness, etc. In the long run, the person may continue to live his normal life without any problem. However, the person may harbour the parasite serving as a carrier of the disease.

(iii) Abscesses: When trophozoites enter blood circulation and other body parts like liver, brain, lungs, spleen, gonads, etc., they cause destruction of tissues leading to abscesses. In the case of liver infection, the liver becomes enlarged, congested and painful to touch. The patient loses his appetite and develops fever. In severe cases, the parasites cause lesions leading to abscesses retarding the metabolic activities. Abscesses of the brain and lung are usually fatal.

Human amoebiasis caused by *Entamoeba histolytica* can be cured by the subcutaneous injection of ematine and by some other oral drugs. Prevention of human amoebiasis is also a matter of personal and public hygiene.

2

PHYLUM PLATYHELMINTHES

General Characters of Platyhelminthes

The Platyhelminthes or Flat-worms are a group of soft bodied, bilateral, usually flattened animals in which there are no true segmentations of the bodies. When the body is short, it may be described as leaf-like; when more elongated, it may be ribbon-like; some have the body comparatively thick. Dorsal, ventral, anterior and posterior regions are well defined in the body. Mouth when present, it is placed generally on the ventral surface. A distinct head is rarely developed.

In Turbellaria, there are cases of having retractile anteior ends. In some Polycladida, a small ventral sucker is developed. In Rhabdocoeles, the anterior and posterior ends are adhesive. In the temnocephalidae there are adhesive tentacles in the anterior end.

In Trematoda there are anterior and posterior suckers. Sometimes adhesive papillae are developed in Trematodes in addition to suckers. The number of suckers is variable in Trematodes. Vibratile cilia are also developed on the body surface of the Trematodes.

In Cestoda the head or Scolex is provided with hooks. The Turbellarians are usually coloured. Pigment is rare in the endoparasitic Trematodes and also Cestodes.

Integument and muscular layers

Platyhelminthes are triplobastic (having epidermis, endodermis and mesodermis in between the two). A peculiar form of connective tissue – parenchyma is developed. This fills up the interstices between the organs. Sometimes the parenchyma consists of distinct large cells with greatly vacuolated protoplasm. There are also parenchyma muscles running through the body for the most part in a dosro-ventral direction.

Alimentary system

An alimentary canal is absent in Cestoda though it is present in Tubellaria and Trematoda. When developed, it consists of a mouth (sometimes surrounded by oral sucker), a muscular pharynx and a branched intestine. The pharynx may be rounded or cylindrical and is capable of eversion and retraction. In Actinodactylella an extensible proboscis is developed which becomes retracted within the opening of the mouth. Unicellular glands open into the pharynx in most cases. In Turbellacia the intestine consists of a pair of simple sac with sometimes lateral diverticula. In Trematoda the intestine is represented with a pair of canals bearing numerous diverticula. In some Polycladida there are minute pores, by means of which certain of the canals open to the exterior. In Trematodes and Rhabdococles there are unicellular glands opening at the junction of pharynx and intestine.

Excretory system

The excretory system consists of the flame cells, excretory capillaries, excretory tubules, and a vesicle opening to the exterior through a pore. In Turbellaria, besides the flame cells, certain glandular cells called paranephrocytes occur in close contact with the excretory units and also help in the removal of excretory products.

Nervous System: Sense organs

Nervous system consists of, in most cases, a brain, placed anteriorly, nerve cords which are laterally situated and nerve branches. The brain consists of numerous nerve fibres. In Tricladida, and some Trematodes, there are three paris of longitudinal nerve-cords running backwards from the brain throughout the body. Several nerve branches may be given off from the longitudinal nerve-cords.

Sensory system consists of tactile cones of Trematodes, sensory cilia and auricular organs of Turbellaria. In other cases, eyes and statocysts are developed. While in polycladids there are numerous eyes, these are practically absent in the digenetic Trematodes and Cestodes. A fully developed eye consists of a cup formed of one or more pigment cells having sensory cells connecting the brain.

Vascular System

It consists of a channel of water-vessels concerning osmoregulatory and excretory function. The water vessels further consist of two longitudinal canals in Tricladida; the canals open to the outside through special branches by a series of pores. However, the water-vessels open to the outside in a variety of ways. In Cestodes, there are usually four longitudinal trunks, which open through a contractile excretory sac at the posterior end of the body. In certain other cases, the longitudinal trunks may become associated with a number of transverse branches, while the main trunks are connected by a ring vessel.

Reproductive System

Sexes are usually united (monocious). Male reproductive

organs consist of paired testes, vas deferentia, a seminal vesicle, ejaculatory duct and penis. The female reproductive organs consist of ovary, oviduct, uterus, vitelline and shell glands with associated ducts.

Fertilization is internal; development is either direct or with one or more larval stages.

DISTINCTIVE CHARACTERS AND CLASSIFICATION OF PLATYHELMINTHES

Characteristics of the Phylum

1. Bilateral symmetry; acoelomata.
2. Triploblastic animals (ectoderm, endoderm and mesoderm—form the body).
3. Body soft, usually depressed, often elongated; no skeletal system; muscle layers well-developed.
4. No true segmentation.
5. Body covered by ciliated epidermis or by cuticle.
6. Some with external suckers or hooks or both.
7. Digestive system is incomplete, or none.
8. Excretory system with many flame cells and excretory canals.
9. No circulatory or respiratory system.
10. Nervous system with internal ganglia or nerve ring, and one or two pairs of longitudinal nerve cords.
11. Sexes usually united; reproductive system of each sex is with gonads, ducts and accessory organs.
12. Fertilization is internal; development is direct or with one or more larval stages.
13. Some are free living while some others are either ectoparasites or endoparasites.

Class I. Turbellaria (Flatworms)

1. Mostly non parasitic, carnivorous also. Free living forms—living in fresh waters, brackish, or salt waters, etc.
2. Usually pigmented, some species brilliantly marked with dorso ventrally flattened body.
3. Head bears eyes, ciliated pits, the "auricles".
4. Epidermis ciliated with many mucous glands and rod-like body, rhabdites.
5. Blind digestive cavity with a protrusible, branched muscular pharynx.
6. A pair of anteriorly placed ganglia gives off lateral nerve cords.
7. Multiplication by fission in some. Reproductive organs show division of lavour, e.g., in the development of a yolk gland, which may have arisen as an over nourished (hypertrophied) ovary. Eggs are usually enclosed in shells or cocoons.

Order 1. *Acoela*: (i) Length to 1 to 4 mm. (ii) Mouth anterior; no digestive cavity (food being digested in the mass of endoderm cells); and no excretory system. (iii) Epidermal nervous system diffused or with a plexus or brain. (iv) Exclusively marine – in the Scandinavian coasts.

Example – *Convoluta, Haplodiscus*, etc.

Order 2. *Rhabdocoela :* (i) Size small, usually elongate and cylindrical or compressed. (ii) Digestive cavity unbranched. (iii) Mostly with Protonephridia and oviducts. (iv) Marine or fresh water. Mostly found in South America and Australia.

Example – *Temnocephala, Microstomum*, etc.

Order 3. *Alloecoela*: (i) Simple, bulbose or plicate pharynx. (ii) Intestine may have short diverticula. (iii) Digestive cavity unbranched. (iv) Protonephridia paired, often with more than one main branches and nephridiopore; Testes numerous, penispapilla usually present. (v) Nervous system consists of 3 or 4 pairs of longitudinal trunks. (vi) Marine or fresh water.

Example – *Monotus, Pseudostomum*, etc.

Order 4. *Tricladida*: (i) Elongate, depressed body, mucous forms a coat of the body. (ii) Mouth midventral, with proboscis; intestine 3-branched. (iii) Testes are fallicular, and ovaries paired; single gonopore. (iv) Yolk glands are present. (v) Aquatic or terrestrial.

Example – *Planaria, Euplanaria*, etc.

Order 5. Polycladida: (i) Mostly of broad flattened form, with tentacles. (ii) Mouth posterior, intestine with many irregular branches. (iii) Nervous system with many radiating nerve cords; eyes numerous. (iv) Ovaries and testes numerous and scattered; no yolk glands; sometimes with a larva (Muller's larva) in development (v) Exclusively marine.

Example – *Thysanozoon, Leptoplana*, etc.

Class II Trematoda (The Flukes)

1. Ecto and endoparasites with a leaf-like body, without an epidermis, rhabdoids or external cilia; mostly unpigmented.
2. Body undivided, covered with a thick cuticle, and provided with one or more suckers on ventral surface and about mouth which lies usually at or near the anterior end.
3. Digestive tract complete, forked usually with two branches and ending blindly.
4. One ovary; genital pore usually anterior to the ventral sucker.

5. Development direct in ectoparasites but indirect in the endoparasites involving alternation of generations.

Order 1. *Monogenea* or *Heterocotylia:* (i) A large posterior ventral adhesive organ of one or several suckers, usually with chitinous hooks. (ii) Mouth anterior, subterminal with 2 ventral suckers near by. (iii) Excretory pores – 2, anterior, dorsal. (iv) Common genital pore lies behind the mouth ; development with ciliated larva or direct ; no intermediate host. (v) Chiefly encto-parasites on cold blooded vertebrates, few in Crustaceans, Coppepoda.

Exan e – *Polystoma, Rajonchotyle, Monocotyle*, etc.

Order 2. *idobothria* or *Aspidocotyla:* (i) Oral sucker wanting; anterior end without paired adhesive structures, no hooks. (ii) With an enormous ventral sucker which is frequently subdivided into lineal series of sucking cups or alveoli. (iii) Mouth terminal or sub terminal; simple and straight gut. (iv) Simple nephridiopore, posterior. (v) Entoparasites with simple life cycle and no alternation of hosts.

Example – Aspidogaster, Cotylogaster, etc.

Order 3. *Digenia or Malacocotylea:* (i) Suckers 2 or 1, one about mouth, the second ventral, if present. (ii) No hooks. (iii) Single excretory pore, posterior, and the genital pore is ventral, between the suckers. Uterus long, many eggs. (iv) Entoparasites with complicated life cycle and several larval stages occupying one or more intermediate hosts.

Example – *Schistosoma* (blood flukes) in China; *Fasciola hepatica* (liver fluck of sheep); *Fasciola buski* (intestinal fluke of man); *Gastrodiscus* (in horse).

Class III. Cestoda (Tapeworms)

1. Body form elongated, tape-like and devoid of epidermis and external cilia.

2. Body clothed with a thick layer of cuticle.
3. Head rounded (scolex) which bears organs of adhesion in the form of hooks (Bothria, Bothridia and Acetabulum).
4. Mouth and digestive tract wholly wanting.
5. In most cases body is made up of a number of segments or Proglottids; separated by transverse constriction; each segment with one or two complete hermophroditic reproductive system.
6. Excretion by flame cells.
7. Nervous system with longitudinal nerve strands and anterior commissures; no special sense organ.
8. All hermophrodites; life history includes a bladder worm (Proscolex) and a tapeworm (Strobila) stage, the former in a vertebrate host or invertebrate host, the latter (with one exception) in a vertebrate.

Sub Class I. *Cestodaria (Monozoa)*

1. Body undivided, no scolex.
2. One set of reproductive organs and ten hooked larva.

Order 1. *Amphilinidea:* (i) Body elongated, flattened, sometimes measuring 30 to 40 cm. (ii) With a protrusible Proboscis and frontal glands at the anterior end, but no definite scolex. (iii) Excretory system in the form of a peripheral network of vessels ; flame cells are arranged in clusters of 20 and 30 and enclosed in special bladders. (iv) Calcareous corpuscles are abundant and there are no cuticular spines. (v) Posterior ovary long ; uterus with 3 limbs.

Example – *Amphilina, Schizochoerus*, etc.

Order 2. *Gyrocotylea:* (i) Body is transversely wrinkled and the margins are frilled. (ii) With a cup-shaped protrusible Proboscis at the anterior end and an adhesive organ (Rosett) at the posterior and surrounding a funnel like depression. (iii) Excretory system with a network partly ciliated vessels to which are attached ordinary flame cells ; nephridiopore is anterior. (iv) No calcareous corpuscles in the parenchyma. (v) Testes numerous; uterine, male and female vaginal pores lie posteriorly.

Example – *Gyrocotyle.*

Sub Class II. *Eucestoda (Merozoa)*

1. Body usually elongated and ribbon-like, segmented or unsegmented; anterior end with a scolex bearing adhesive organs.

2. More than one set of genital organs; larva six-hooked.

Order 1. *Tetraphyllidea (Phyllobothrioidea):* (i) Body segmented with a scolex bearing 4 bothridia, often armed with hooks or subdivided by septa. (ii) Found in the intestine of Elasmobranch fishes.

Example – *Phyllobothrium.*

Order 2. *Lecanicephaloidea:* (i) Scolex with four bothridia. (ii) Numerous yolk glands; parasitic in the Elasmobranch fishes.

Example – *Lecanicephalum.*

Order 3. *Protocephaloidea:* (i) Scolex armed or unarmed with minute spines and with 4 sessile suckers devoid of areole or accessory suckers; and an apical sucker or glandular organ. (ii) Dorsal testes with the inner longitudinal muscles or disposed in definite lateral fields; lateral diverticula in the uterus. (iii) Parasitic in the intestine of fresh water fishes, amphibian, reptiles.

Example – *Proteocephalus.*

Order 4. *Diphyllidea:* (i) Scolex contains powerful muscular rostellum, and armed with 2 bothridia and a spiny head stalk. (ii) Testes few and occupy the anterior half of the proglottid; Follicular yolk glands occur as lateral strands. (iii) Larval stages in marine molluscs and crustacea, adult in the intestine of Selachians.

Example – *Echinobothrium*.

Order 5. Tetrarhynchoidea or Trypanorhyncha: (i) Segmented with a scolex of 4 eversible proboscides. (ii) Vagina and sperm duct do not cross and the yolk glands enclose other reproductive parts. (iii) Found in the gut of fishes.

Example – *Tetrarhynchus*.

Order 6. *Bothriocephaloidea or Pseudophyllidea:* (i) Typically with a scolex and provided with 2 to 6 bothridia, sometimes without adhesive organs. (ii) Uterus persistent either coiled or lobulated. (iii) Found in the Crustaceans and Mammals.

Example – *Bothriocephalus, Ligula*, etc.

Order 7. *Nippotaeudea:* (i) Scolex having an apical sucker but no other adhesive structures. (ii) Single yolk gland; excretory system with numerous longitudinal vessels.

Example – *Nippotaenia*.

Order 8. *Taenoidea or Cyclophyllidea:* (i) Scolex with 4 acebula (suckers), with apical rostellum. (ii) Single compact york gland. (iii) Excretory system having 4 longitudinal canals.

Example – *Taenia solum* (parasite in the human intestine),

T. saginata (beef worm); *T. echinococus* (within small intestine of dog).

Order 9. Aporida: (i) 4 acetabula and armed rostellum. (ii)

No sperm duct, pores and distinct yolk glands; Follicle of germovitellaria become uterine capsules.

Example – Gastrotaenia.

STRUCTURE AND LIFE HISTORY OF

Fasciola Hepatica (liver fluke)

The Liver fluke (*Fasciola hepatica*) is an internal parasite (endoparasite) living in the bile duct of the liver of sheep, goat or cattle (but rarely in man). It causes a destructive disease called "liver rot" or "Fascioliasis". When a sheep suffers from the disease, its wool falls off and on the skin, there appear numerous swellings containing some sort of fluid.

1. Structure

The body is soft, flattened and leaf-like, about 25 mm long and about 12 to 13 mm at its broader part. The anterior side is broad and the posterior portion is tapering. The anterior end forms a cone-like head process at the tip of which is a circular mouth, surrounded by a cup-shaped disc, the oral sucker. A little away from the head process and on the ventral side, there is another saucer-shaped posterior or ventral sucker which serves only for attachement. Approximately in the middle line between the anterior and posterior sucker is the genital opening or gonopore (a common aperture for both male and female organs). At the extreme posterior median end is an opening called the excretory pore. On the dorsal

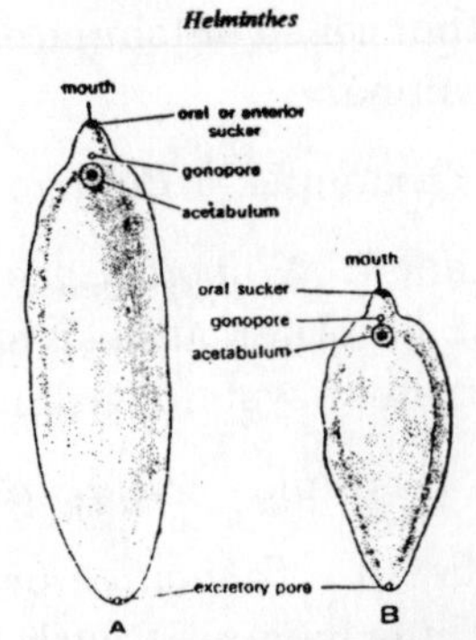

Fig. 5-2. Liver flukes. External features in ventral view. A—*Fasciola gigantica* (=*F. indica*). B—*Fasciola hepatica*.

surface, at about a third of the length from the anterior median side is located the dorsal pore or opening of Laurer's canal.

2. **Body wall**

The surface of the body is invested by a thin layer of cuticle possessing backwardly directed spinules which help the animal to hold firmly to the substrate with which it is in contact. Below the cuticle there is a well developed musculature consisting of an outer circular layer and an inner longitudinal layer. Epidermis is wanting as a separate cell layer. There is no body cavity, and the space between the body wall and the internal organs is occupied by a mass of tissue called parenchyma.

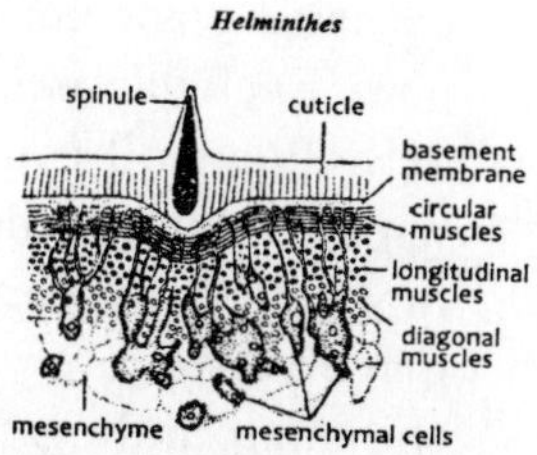

Fascioloa helpatica. T.S. of the body wall under light microscope

3. The alimentary system comprises of the mouth, pharynx, oesophagus and a branched intestine. There are two main branches of the intestine – a right and a left branch, each running to the posterior end of the body, one on either side of the middle region. These two main branches give off secondary and tertiary branches (diverticula) on their outer sides. The intestinal branches end blindly. There is no anal aperture.

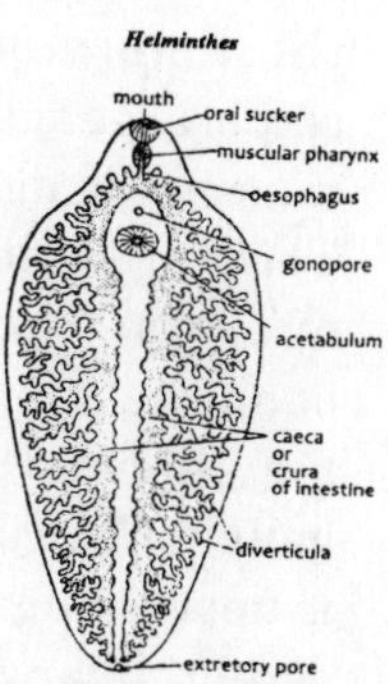

Fasciola hepatica. Digestive system.

4. There is no circulatory or respiratory system. The parasite performs anaerobic respiration.

5. The excretory system consists of numerous branched tubules ending in flame cells. The tubules on the other hand all unite in the middle line to form a longitudinal or median excretory canal which opens to the exterior posteriorly by the excretory pore. Each flame cell is a minute vesicle or hollow cell containing a few cilia which keep moving like the flickering movement of a candle flame. The cilia drive out excretory fluid, collected from the surrounding cells into the minute excretory tubule and then to the ducts that open to the main excretory canal.

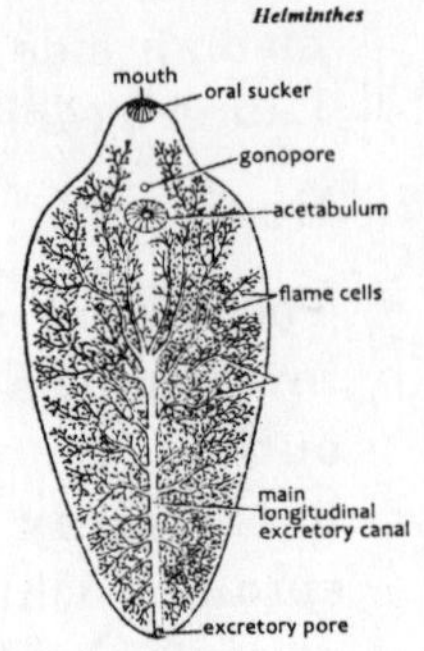

Fasciola hepatica. Excretory system.

6. The nervous system consists of a nerve ring surrounding the oesophagus, a pair of cerebral ganglia, a single ventral ganglion, two nerve cords and their nerve branches. The cerebral ganglia are prominent masses of nervous material. They are lateral in position, one on the side of the oesophagus. The cerebral ganglia and the midventral ganglion are joined together by a nerve ring that encircles the oesophagus. From the cerebral ganglia nerves are given off to the head process and to the hinder end of the body. Of the nerves arising from the cerebral ganglia, there are two large nerve cords. These run backwards to the hinder end of the body and send off smaller other nerve branches in this region.

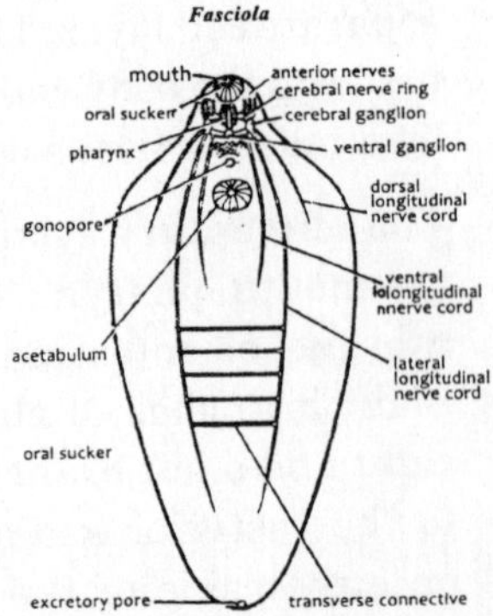

Fasciola hepatica. Nervous system.

In the adult Liver fluke, there are no sense organs; however, the larva has certain eye spots to detect light and darkness.

7. The animal is hermaphrodite with well developed male and female organs. The male reproductive system consists of a pair of testes, a pair of vasa deferentia, a seminal vesicle, an ejaculatory duct, a common genitalatrium opening to the exterior by the gonopore. The two testes are situated one behind the other in the middle part of the body. Each testis is a large, much branched tube opening into a duct – the vasa deferens. The two vasa deferentia run forwards and join a sac-like seminal vesicle which stores the spermatozoa. From this seminal vesicle arises a twisted tube, the ejaculatory duct which passes forwards and continues into a muscular penis (cirrus) surrounded by a protrusible sac, the cirrus sac. The penis opens to the outside by the male pore situated into the genital atrium.

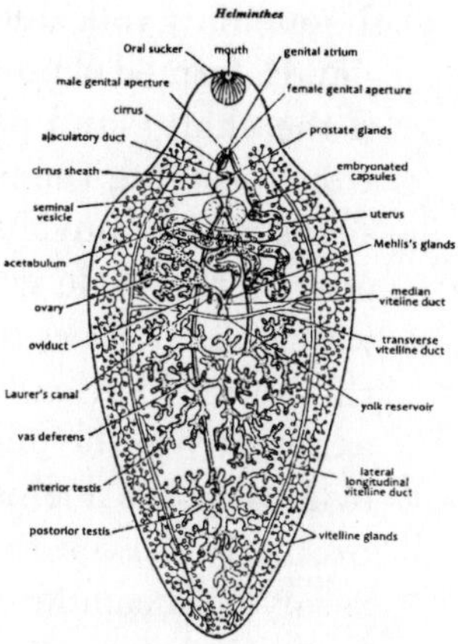

Fasciola Reproductive system in ventral view.

The female reproductive system consists of a single ovary situated on the right side of the body in front of the anterior testis. It is much branched and the branches join to form a narrow oviduct. The oviduct runs forwards to the middle line and joins the median viteline duct. Occurring along the sides of the animal are numerous, small vesicles, the yolk glands or vitellarium. Each yolk gland secretes yolk and has its own duct; the various yolk ducts lead into an anterior and a posterior longitudinal vitelline duct one on either side. The longitudinal ducts unite to form a right and a left transverse duct. The two transverse ducts combine to form a median yolk duct which joins the oviduct. This junction of the oviduct with the median yolk duct is surrounded by a group of shell glands, formed of minute unicellular glands. The secretion of the shell gland passes into the oviduct. The oviduct, after receiving the median yolk duct, proceeds forwards as the broad, convoluted, uterus which opens outside very near to the male gonopore. The part of the uterus into which the oviduct opens is sometimes distinguished as the ootype. A canal called the Laurer's canal opens externally in the mid-dorsal line, and internally into the oviduct where it joins the median yolk duct. Laurer's canal probably serves for the entrance of the spermatozoa received in copulation from another Liver fluke.

Life History

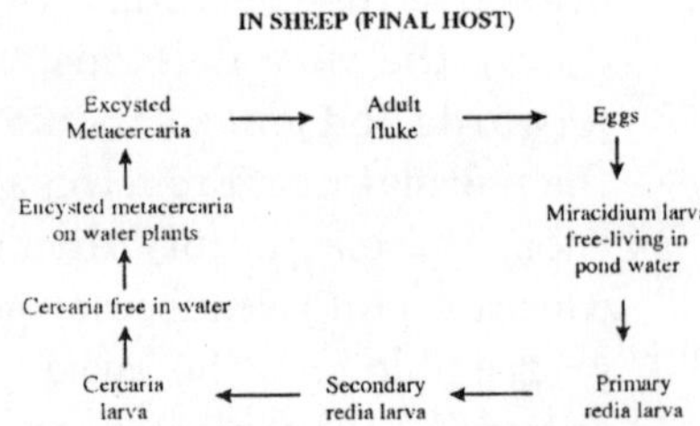

A Liver fluke may produce as many as 500,000 eggs at a time in the bile ducts of the sheep. They are carried to the intestine and passed outside with the faeces of the sheep. If the eggs are laid on moist and warmer soil, they develop usually within 9 days.

From the eggs, there hatch out multicellular ciliated larvae called Miracidia. The Miracidium is very small in size and is about 1.5 mm in length. The body form is conical with an outer layer of hexagonal cells arranged in five bands. All over its body there are cilia for locomotion. At the anterior broader end there is a triangular boring papilla. At the back of the boring papilla is a nerve which is movable ganglion with a pair of crossed eye-spots for recognizing light and darkness. There is a small gut and a pair of flame cells which open to the exterior by minute pores. The interior of the body cavity is filled with loose parenchymatic tissue and few reproductive cells.

Helminthes

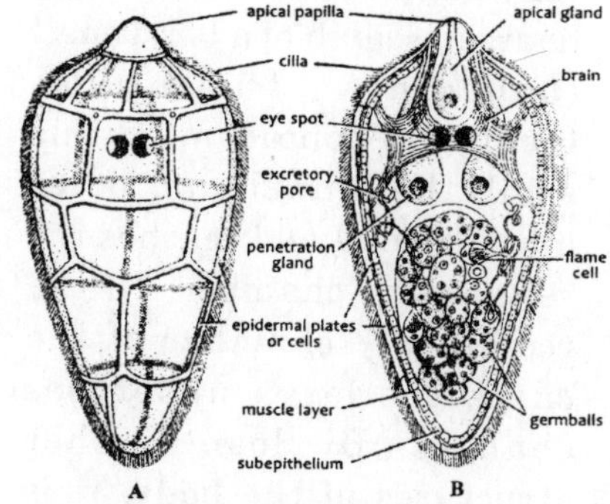

Fasciola hepatica. Miracidium larva.
A - External view showing epidermal plates.
B - Internal structure.

Helminthes

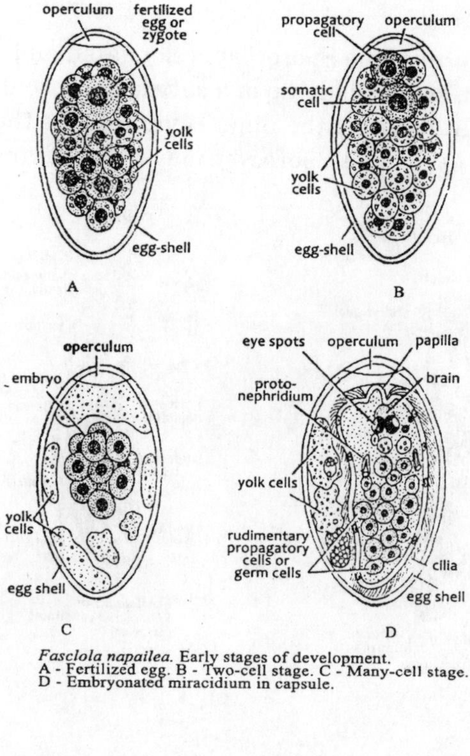

Fasclola napailea. Early stages of development.
A - Fertilized egg. B - Two-cell stage. C - Many-cell stage.
D - Embryonated miracidium in capsule.

With all these characters, Miracidium swims for about 8 hours in search of a pond snail (*Limnaea truncutula*). Coming in contact with this snail, the Miracidium bores into the snail and reaches the respiratory chamber. In the respiratory chamber of the snail, the larva undergoes changes. By losing other structures of the body, it is transformed into the Sporocyst.

Helminthes

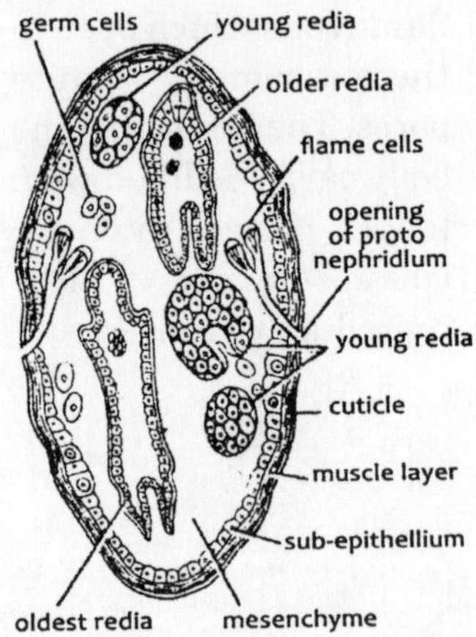

Fasciola hepatica

The body form of the Sporocyst is elongated and sac-like. The wall of the sac is made up of a cuticle, a muscular layer and an epithelium lining the inner cavity. Inside the cavity, are few germ cells. The Sporocyst may divide transversely

Fasciola

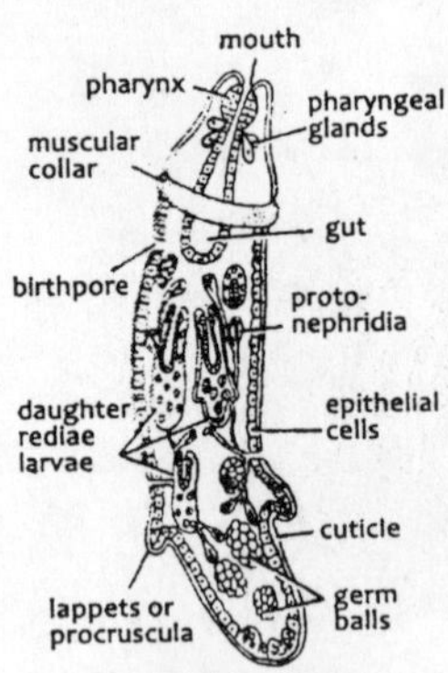

Fasciola hepatica.
Redia larva containing daughter rediae.

Fasciola

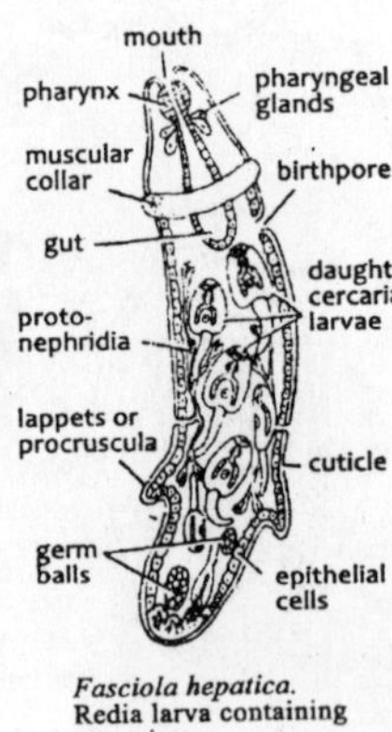

Fasciola hepatica.
Redia larva containing cercariae.

and produce other Sporocysts. The germ cells within the Sporocyst can develop by parthenogenesis into another kind of larva called Redia. A Sporocyst produces four to eight Redia.

The Redia has a cylindrical body form. It has a mouth and a short gut, a pair of blunt projections near the posterior end and a minute birth-pore. When the Radiae attain adult size, they break the wall of the Sporocyst and migrate to the host's liver. In the meanwhile, the germ cells of the Redia develop, by pathenogenesis, into the new type of larva called Cercaria.

Helminthes

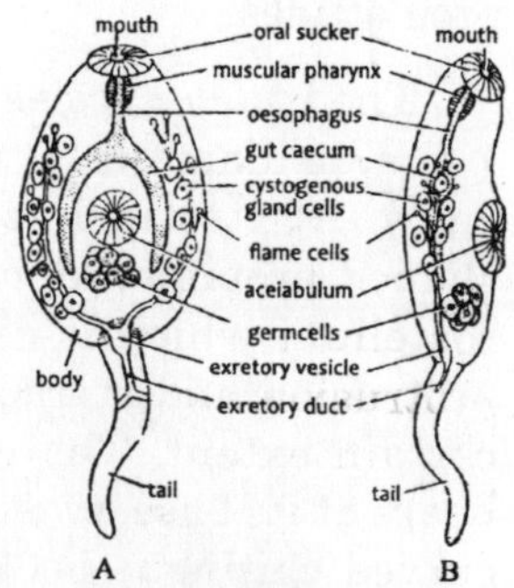

Fasciola hepatica. Cercaria larva.
A- Ventral view. B - view.

Cercaria resembles to the adult Liver fluke. It has a disc-like body provided with anterior and ventral suckers. The alimentary canal is developed with mouth, pharynx and U-shaped intestine with no anus. A nerve cord surrounds the pharynx. The interior of the body is occupied by parenchyma and few gland cells. There is a long tail for locomotion.

The Cercaria escapes through the birth-pore and settles down on a grass blade and undergoes further changes. It loses its tail and secretes a cyst round itself and is then transformed into Metacercaria.

If the Metacercaria is taken by the sheep along with the grass, the cyst wall is dissolved and the larva burrows through the intestinal wall into the body cavity and finally migrates the liver where it feeds and grows to become the adult Liver fluke.

STRUCTURE AND LIFE HISTORY OF TAPEWORM (*Taenia Solium*)

The common Indian pork Tape worm (*Taenia solium*) is found mostly in those places, where pork is eaten as an item of food. It is an endoparasite, living in the intestine of man.

It is a digenetic form. Man is the definitive or primary host and pig is the secondary host.

Structure

The parasite is whitish in colour. Its body is flat, elongated and ribbon-like. It may grow upto 2-3 metres in length and 6 mm. in breadth. The body is distinguishable into (a) head or scolex, (b) neck and (c) a long chain of segments – strobila or proglottides.

The head or scolex is a pear shaped anterior part, about 1 mm. in diameter. It is devoid of mouth. It is muscular and rigid. At its tip lies a cone-shaped prominence, called the rostellum which is capable of protrusion and retraction to a certain extent. The rostellum bears at its base, two circlets of curved chitinous hooks (about 25 in number). Behind the rostellum, the scolex is provided round the circumference, four adhesive suckers known as acetabula.

Taenia solium

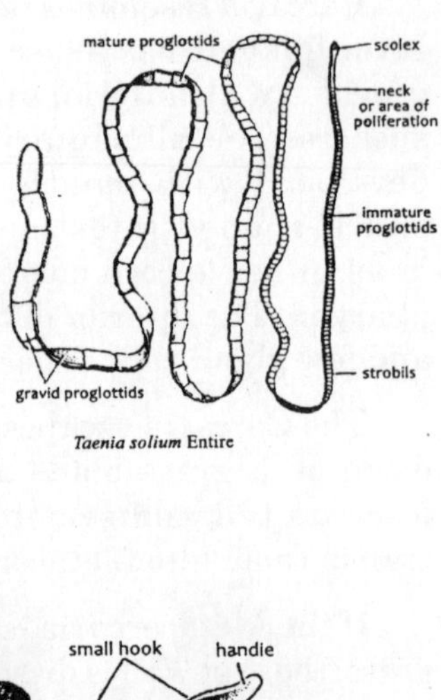

Taenia solium Entire

Taenia solium

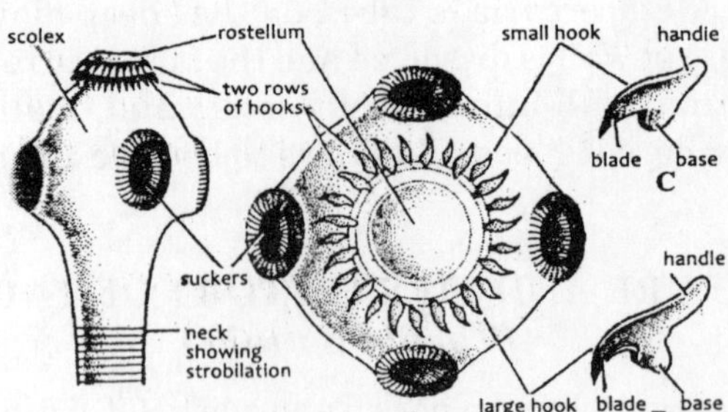

Taenia solium.
A-Scolex with neck in side view. B - Scolex in enface view. C - Small hook. D - Large hook.

Neck is the "growing region", just behind the scolex. It buds off more and more new segments, called proglottides.

The proglottides may be of three kinds :

(a) *Immature proglottides*: These are broader proglottides of the anterior region. They contain only developing reproductive organs and excretory ducts.

(b) *Mature proglottides*: They are the proglottides of the middle region. Each of them is rectangular in outline and protected outside by a thick cuticle which is perforated at intervals by canals. Below the cuticle is a layer of outer circular and inner longitudinal muscles. In addition to these muscles, there are also the dorsoventral, transverse and longitudinal muscle fibres. Towards each lateral region of the proglottides, are found lateral nerves and excretory canals. Transverse excretory canals connect the lateral canals. Male and female reproductive organs occupy the major portion of the proglottides.

(c) *Gravid proglottides*: They are proglottides of the hinder region of the strobila. They are longer than broad. Each of them possesses a much branched uterus lodging numerous eggs. The other reproductive parts are degenerating.

Internal Structures

1. Tapeworm has no digestive or respiratory organs. General surface of the body serves the function of absorbing food and oxygen from the host's body.
2. Excretory system consists of flame cells and their connecting tubes.
3. Nervous system consists of ganglionated band of nervous material in the head. From this, nerves are given off to the suckers and lateral sides of the proglottides.

4. The animal is hermaphrodite. The male reproductive organs mature at about 400th segments. The organ consists of much lobed testes having efferent duct, vas deferens, cirrus and male gonopore. The female reproductive organ comprises of paired ovaries with their ducts, uterus, yolk and shell glands, receptaculum seminis, vagina and the female gonopore.

Life History

Fertilization may take place usually between different proglottides of the same worm or within a proglottid. Many eggs may be fertilized at one time in the lower part of the oviduct and on their way to the uterus they receive a supply of yolk from the vitelline glands. These glands also secrete a thin shell around each egg. In the uterus development begins and the first division of the egg-cell separates a large vitelline cell from an embryonic cell. The embryonic cell divides forming a solid ball of cells, the morula. The morula has an outer layer of a few macromeres and an inner mass of smaller cells. The cells forming the outer layer of the morula then secrete a thick striated coat formed of chitin. This is the embryophore. The inner mass of cells develop into six-hooked embryo, the hexacanth. The hexacanth together with the thin egg shell, and embryophore is known as onchosphere. The ripe proglottides at the extreme end of the strobila contain embryos in this stage of

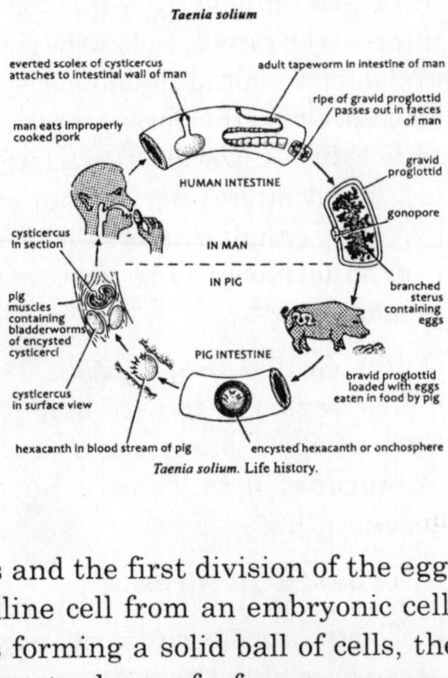

Taenia solium. Life history.

development. The terminal proglottides drop off from the strobila and pass out with the faeces of the host.

When swallowed by the omnivorous pig along with the faeces, the egg shell and the embryophore dissolve by the action of the digestive juices of the pig. The hexacanth embryo bores through the wall of the gut by means of the hooks and enters into the blood stream. Blood carries the hexacanth to the voluntary muscles of tongue, limbs, etc.

Within the muscles the embryo loses its hooks and develops into a large cyst with a large cavity. The wall of the cyst is pushed inward at one point and the impushed wall forms a proscolex. The scolex is then evaginated, and the cyst with the scolex looks like a bladder and is called the 'bladder worm stage' or cysticercus. The pig's flesh infected with the cysticercus bears brown spots and is called measly pork.

If the ill cooked or uncooked measly pork is eaten by man, the bladder is lost and the scolex fixes itself to the wall of the intestine of man. It develops series of proglottides to reach the adult stage. Within 2-3 months time, the worm attains sexual maturity and survives for about 25 years.

Taenia solium

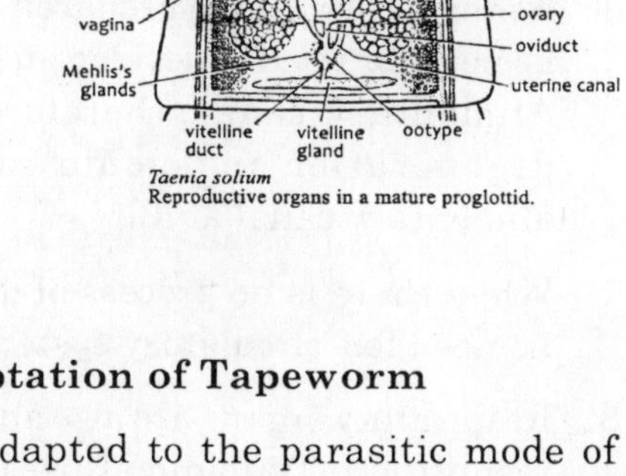

Taenia solium
Reproductive organs in a mature proglottid.

Parasitic Adaptation of Tapeworm

Tapeworm is highly adapted to the parasitic mode of existence. It has undergone modifications of structures and functions which are in the nature of adaptations to parasitic life. Some organs are simplified, some complicated, while some others are degenerated.

Modifications

1. The body of the Tapeworm is flattened and ribbon-like with head, neck and proglottides.
2. The head is armed with four suckers and two whorls of hooks around the rostellum to ensure better attachment to the intestinal wall of the host.
3. The thick cuticle of the parasite is surrounded by the mucous from the intestinal wall of the host. The parasite stimulates the gut of the host to secrete this mucous. Added to this, the parasite also secretes an "antienzyme" to nutralise the action of the digestive enzyme of the host. Thus the parasite protects itself from destruction by the host.
4. The entire outer surface of the parasite in such a location usually serves as a absorptive surface. This is not only the case in the adult stage but also the larval forms.
5. The locomotory organs and skeleton are absent because the parasite is satisfied with its hunger and protection within the host.
6. As the parasite gets prepared food from the host, there is no need for itself to develop an elaborate digestive system. Alimentary canal, therefore, reaches the height of degeneration in Taenia in which there exists no alimentary canal at all.
7. When there is no process of digestion evidently there is no need for, circulatory system.
8. Respiratory organs are also absent. It performs anaerobic respiration. The animal uses glycogen as the source of energy and carbon dioxide and organic acids are liberated.

9. Presence of flame cells in the excretory system facilitates collection of the nitrogenous wastes from the nook and corner of the body.
10. The expenditure for vital energy is minimum.
11. The internal osmotic pressure of tissues of the parasite is nearly the same as that of the surrounding fluid of the host. This is a good case of biochemical adaptation of the parasite.
12. The adult parasite is practically a sedentary form and its environment is rarely subjected to any appreciable changes so that the importance of the nervous system and sensory organs is decreased. So, its nervous system suffers degeneracy, consisting of only a pair of cerebral ganglia and a pair of longitudinal nerve cords.
13. If the parasite can ensure that its eggs and offsprings reach another host, then only the parasite can maintain the continuity of its own race after the death of the host. This is made possible by complicating its life cycle. First of all, the parasite has become hermaphrodite. By this means, it has substituted self-fertilization for cross-fertilization to a great extent. Every segment of the body is a self-sufficient unit with male and female reproductive organs. These organs are renewed by budding of new segments of the body within the host. The whole reproductive system in a segment is capable of producing huge number of eggs per day.
14. The eggs become enclosed in a thick resistant shell within which they can remain viable for a long time.
15. The structures of the larval forms are well suited to the entry of the new host. The Cysticercus larva acquires the bladder so that when the larva inside is exposed to the action of the digestive juices of the host, the bladder is

dissolved but the inverted scolex within it survives, thus resuming further development.

PARASITISM, ORIGIN OF PARASITISM AND PARASITIC ADAPTATION

Parasitism is the association of two organisms of same as different species of which one lives at the cost of the other. The one which victimises the other living organism to get nourishment is called the parasite, and the other which is a victim is known as the host. The parasites are biologically and economically intimately connected with the hosts. The hosts offer resistence by producing antibodies and become tolerant to the presence of the parasites.

The parasitism may be accidental, facultative or obligatory. In accidental case of parasitism a free living animal, if taken accidentally by the host, is capable of leading a parasitic existence for a certain period. In facultative type, the organism can live both as free living or parasitic. In obligatory mode, the parasites must live on some other suitable organism, otherwise they cannot exist.

Origin of Parasitism

The parasitic mode of life is a secondary state of affairs. It has arisen from free living way of life. It began as a chance contact of one organism with another. Smaller animals occassionally look for temporary shelter on the body of some larger animals. When the temporary shelter is repeatedly used, the association of the smaller and larger forms become very intimate. The larger organism being merely a vehicle transporting the smaller organism from one free living feeding ground to another. Sooner or later, the guest comes to pertake of the food procured by the host, becoming more and more dependent on such food and gradually changing from ectoparasitic to endoparasitic condition. In this way parasitism originated.

In some cases, the guest begin to consume the tissues of the host first in a very superficial way. But later, by use of suctorial organ developed for such feeding or by actual penetration into and residence in the tissues of the host, resulting in a very degenerate and dependent existence for the parasite. Such forms are those which are able to live with the blood or lymph system or which have residence in the muscular tissues or attached to the peritonerm or pleura are usually most highly modified, and are therefore, regarded as the oldest parasitic species, other species which live in the mouth or bladder of the host or on the surface of the body are frequently very young in parasitism.

PARASITIC ADAPTATION
(Adaptation of Helminthes to a parasitic mode of life)

The change from a fortuitous free living existence to one in which protection from enemies and a good supply of food is granted, has brought about a profound modifications in the structure and physiology of the helminth parasites.

Endoparasites are more modified then the ectoparasites. This is because the endoparasites live in situations where they are surrounded by food material which is continuously available.

ADAPTATIONS

I. Morphological Loss of Structure

The trophic organs that is, the organs concerned with nutrition of parasites are reduced or entirely lacking when compared with those of their free-living allies. As the parasites live in the digested or semidigested food of the host, their organs of locomotion and nutrition have become simplified. A mouth and incomplete digestive tract is retained by flukes but these are lost in Tapeworms. Trematodes have a simple suctorial pharynx by means of which nourishment is drawn from the tissues of the host. In Redia stage the organ is simplified, and in Sporocyst stage it is completely eliminated. Tapeworms absorb through their cuticle the food that is digested by the host.

Modifications

The body of the parasite is covered by a thick cuticle which is often raised into minute spines or scales. The cuticle contains highly insoluble proteins. The entire outer surface of the parasite serves as an absorptive surface. This is not only the case in larval stages but also the adult flukes in the hepatic portal system and the various species of liver flukes in the bile tracts are also provided with absorptive integument.

The outer integument covering the body of the helminth parasites residing in the intestinal lumen have become adapted for protecting the organism:

(1) from the digestive action of intestinal juices;

(2) from the absorptive action of food and roughage passing through the gut, and

(3) the integument has also been modified for attachment.

(1) Protection against intestinal juice:

(i) In the case of larval liver flukes which have to pass through the stomach in order to reach the bile passage for further development, a cyst capsule is provided as a protection against gastric juice.

(ii) Certain Amphistome species (in ruminants) and Gnathostome species live attached to the stomach wall. In consequence, they are provided with chitin-like substance of specially impermeable character.

(2) Protection against abrasion:

(i) Many Trematodes living in the intestinal tracts are provided with spinous integument to guard against the abrasive action of the food and roughage passing through the gut. The spines may be acicular (needle shaped) or denate and rooted into the subintegumental layer.

(ii) *Clonorchis sinensis,* an intestinal parasite, possesses spinous integument in larval period. The adults are entirely without spines.

(3) For the purpose of attachment to the intestinal wall of the host:

(i) Acetabulum or Sucking organ : It is found in all of the flatworms which parasitize men.

(a) In the flukes, they consist of two suckers on the ventral side of the body, one anterior and the other more or less posterior in position.

(b) In the case of human Tapeworm, they consist of either sucking tongue or grooves or 4 cups at the cephalic end of the worm.

(ii) Hooks: In some of the Tapeworms and many Nematodes which are parasites of the human intestine, hooks are situated in or around the anterior end to assist in attachment.

(a) In *Taenia solium* hooks are arranged in a circlet at the base of the rostellum.

(b) In Dog-Tapeworms, they occur in several rows around the proboscis-like organ at the anterior most part of the body, which may be everted or inverted.

(c) The Hook-worm has a series of upper teeth within its buccal capsule.

(d) Triodontophorous has a buccal armature of tooth-like structure directed anteriorly which serves both for the tissue-aberration and for anchoring of the parasite.

Development of secretory glands

In some helminthes there have been developed in the vicinity of the mouth secretory glands which serve in –

(i) establishing the worm in a favourable habitat, or

(ii) aiding supply of food to the worm.

(iii) In the Trematodes these glands are more common in the Cercaria stage, and serve the purpose of penetrating the outer tissues of the host by the elaboration of histolytic substance.

(iv) In the case of hookworms there are glands in the region of the buccal opening which helps in getting a continuous supply of uncoagulated blood and mucous cells for their food. The glands are supposed to have anticoagulating action.

Reproductive organs

The most conspicuous increase in organs and tissues of the helminthes as a group is that of the reproductive system. Both Platyhelminthes and Nemathelminthes have a large part of the body occupied by these organs and their products.

In Tapeworm instead of a single body unit there are multiple segments – Proglottides, each one sexually complete itself. Thus a single worm may produce fertilized ova numbering into thousands daily. This enormous production of offspring is necessary because the chances of the offspring reaching the main host are small in many cases. The entire life cycle may involve one to three intermediate hosts with many larval stages.

Nervous and Excretory systems

The nervous and excretory systems in the case of parasites have been little altered in the adaptation of the organisms to a parasitic mode of existence. Rudimentary eyes are however, present in the free living miracidium larva of liver fluke.

II. Physiological Adaptations

The parasites have undergone certain physiological adaptations as in the following:

1. In order to prevent the disturbing exchange of water, the osmotic pressure of the body fluids of parasites has become the same as that of the host.
2. In most internal habitat of the parasite the oxygen content is very low. So the parasites become modified to obtain oxygen by anaerobic respiration. They obtain their energy by anoxybiotic process utilizing glycogen as the source of energy.
3. The parasites can secrete antienzymes to neutralize the digestive juices of the host.
4. Many Castodes are able to stimulate the gut of the host to secrete mucus which surrounds the tapeworm as a protective envelope.

Conclusion

Viewing the group of parasitic helminthes a whole with a successive stage of adaptation, which they have undergone and are undergoing, and as a result on the one hand, the specialization of organs for resistence, attachment, reproduction, and on the other, the reduction in the organs of locomotion and alimentation, one is able to appreciate how profound must have been the alteration from a free living existence and how dependent the parasite is upon the host when once it has become so adapted.

3

PHYLUM NEMATHELMINTHES (ASCHELMINTHES)

Characters

1. The Nemathelminthes (Aschelminthes) are predominantly aquatic but some are terrestrial. They are free-living or parasitic. They are small sized animals, a few attain notable lengths.
2. The body is vermiform either cylindrical or flattened without definite head.
3. The body is covered with tough resistant cuticle which may be ringed or covered by spines, scoles or bristles, etc.
4. Below the cuticle lies the epidermis which is syncytial in most cases, and the musculature lying beneath it is not arranged into definite layers.
5. They are pseudococlomate with unsegmented or superficially segmented body.
6. The alimentary canal is straight or curved without definite muscular wall. The pharynx is highly differentiated. The anus is situated posteriorly.
7. In some, flame cells with excretory canals are present.

8. There are no respiratory and circulatory systems.
9. The nervous system is well developed. It consists of an anterior brain mass, a circumenteric nerve ring and ventral and lateral nerves.
10. The reproductive system is simple, and the sexes are separate.

Classification

Class I. Rotifera

1. The rotifers or wheel animalucles are microscopic, ranging between 0.04 to 2mm in length. They are mostly free living.
2. Body is elongated and divisible into an anterior head region bearing ciliary apparatus, the corona, an elongate trunk and the slender terminal region, the tail or foot. Some forms are sac-like while some others are either broad or slender.
3. The body is covered by cuticle. In some cases the cuticle is thickened in the trunk region forming a hard encasement, the lorica.
4. Mouth is anterior and more or less ventral in position. Pharynx contains a chitinous masticatory apparatus. Anus is dorsal at the junction of the trunk with the tail.
5. There is a spaceous body cavity devoid of epithelial lining.
6. The excretory system consists of a pair of flame bulb.
7. Central nervous system consists of a single dorsal ganglion; in some cases a smaller ventral or sub oesophageal ganglia also occurs.
8. Sensory organs are the eyes located in the brain region, lateral region and apical region.

9. Sexes are separate. Male organs are degenerating and smaller than the female organs.

   Example – *Scison, Philodina, Brachionus*, etc.

Class II Gastrotricha

1. They are free living, found commonly among the algae, at the bottom of the fresh and salt waters.
2. The body is elongate with flat ventral surface, unsegmented, without corona, ciliation being restricted to limited parts. The posterior end is forked.
3. The cuticle is often provided with spines, plates, scales.
4. Alimentary canal is straight with mouth at the anterior end and anus at the posterior end.
5. The body cavity is a pseudocoelom.
6. There are two protonephridia which are coiled and unbranched, each is provided with flame cells, opening to the exterior.
7. A ventral nerve cord is present.
8. They are mostly hermaphrodite; parthenogenetic females are also found. Development is direct.

   Example – *Chaetonotus, Cephalodasys*, etc.

Class III Echinodera (Kinorhyncha)

1. They are minute marine worms of cylindrical form having a segmented body with spines.
2. The elongated body is divisible into a spherical head connected by a neck to the jointed trunk. The head is retractile.
3. The body wall consists of a cuticle below which lies a syncytial epidermis. Both transverse ring muscles and longitudinal muscles are present.

4. In the alimentary canal, the mouth opens into a conical buccal cavity which leads into the pharynx. The anterior end of the pharynx bears a cuticular ring having teeth-like structures and projecting into the buccal cavity. The pharynx is followed by a slender oesophagus, stomach – intestine and the hind gut. There are two or more pancreatic glands or salivary glands at the junction of the stomach and oesophagus. The anus opens at the posterior terminal end.
5. Pseuodcoel is filled with a fluid containing amoebocytes.
6. Excretory system consists of a pair of nephridia each with a flame bulb and opening at the nephridiopore.
7. In the nervous system, there is a circumenteric ring surrounding the base of the mouth. A ventral ganglionated nerve cord arises from the nerve ring.
8. Sensory organs consist of eyes and sensory bristles.
9. The sexes are separate. The gonads are a pair of sac-like syncytial bodies opening separately on the 13th zonite.

   Example – *Echinoderes, Centroderes*, etc.

Class IV Priapulida

1. They are cylindrical worm-like marine animals.
2. The body consists of an anterior region, the presoma, a large trunk and one or two warty appendages at the posterior end.
3. The body wall consists of an outer cuticle and inner epidermis below which lies the circular and longitudinal muscles.
4. The protonephridia are solenocytic.
5. The mouth is terminal and surrounded by a number of strong spines. The mouth opens into the muscular

pharynx which is followed by the midgut or intestine. The anus is at the posterior end of the body.

6. The nervous system comprises of a circumentric ring surrounding anterior end of the pharynx, and a mid ventral cord lying in the epidermis.

7. They are dioecious.

   Example – *Priapulus.*

Class V Nematoda

1. The body is bilaterally symmetrical, non-segmented, cylindrical and pointed at both ends.
2. The body wall is covered externally by a cuticle and marked by four longitudinal lines. Below the cuticle lies the epidermis which is formed of a single layer of cell syncytium. Muscle layer consists of only longitudinal fibres with club-shaped processes.
3. They are triploblastic and coelomate.
4. Digestive system is complete with mouth at the terminal anterior end and anus on the ventral surface, close to the posterior end.
5. The circulatory and respiratory systems are absent.
6. The excretory organs are in the form of a fine longitudinal canal running through the body wall on each side and opening by a minute mid-ventral pore, behind the mouth.
7. The nervous system consists of a nerve ring encircling the oesophagus, and longitudinal nerve cords.
8. The sexes are separate. The males are smaller than the females. The posterior end of the male is curved or expanded into a cudal bursa. The gonads are continuous with reproductive ducts, single or double in male, paired or single in female.

9. The eggs are microscopic and are covered by a chitinous shell. Development is direct, immature forms undergo moults to become the adult forms.

   Example – *Ascars, Trichinella spiralis, Ancylostoma, Wuchereria, Oxyuris*, etc.

Class VI Nematomorpha (Gordiacea)

1. The body is cylindrical and long with a rounded anterior end and a swollen and coiled posterior end.
2. They are sexual as adults but parasite and asexual as juveniles, and free living.
3. The body cavity is filled with parenchyma.
4. The digestive tube is degenerate.
5. They have a single nerve cord and separate gonads. The genital ducts open into the cloaca.
6. The larvae are exceedingly small and they possess the proboscis bearing hooks and spines. They bore into small animals and encyst.

   Example – *Gordius, Nectonoma*, etc.

STRUCTURE AND LIFE HISTORY OF ASCARIS (ROUND WORM – Phylum Nemathelminthes)

Ascaris is a nematode parasite which lives in the intestine of man, more frequently among children. Usually it is very troublesome, though not so dangerous. It feeds on partially or wholly digested food of the host.

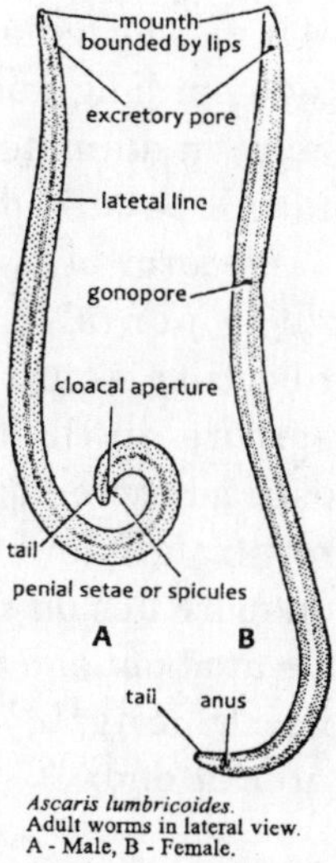

Ascaris lumbricoides.
Adult worms in lateral view.
A - Male, B - Female.

External Characters

Round worm has an elongated, cylindrical body, pointed at both ends. It is not segmented. The wrinkled nature of the outer covering, gives a segmental appearance to the animal. The covering cuticle is thick, yellowish brown in colour. Along the entire length of the body there are four longitudinal streaks – one dorsal, one ventral and two laterals. The mouth is anterior and terminal, markedly constricted off, and is bounded by three lips – one median and dorsal and two ventro-lateral. The lips bear sensory papillae. At about 2 mm. behind the anterior end on the ventral side is the excretory pore. The anus is a transverse opening on the ventral side close to the posterior end. Anus performs double functions of elimination of faeces and removing the genital products, and so it may be called cloacal aperture.

The male Ascaris is smaller than the female and it measures 15-25 cm. long with a maximum diameter of 3-4 mm. The female is about 25-40 cm long and about 6-8 mm. in diameter. In male the tail is curved ventrally in the form of a hook with two penial spicules projecting from the cloacal aperture. In the female the posterior extremity is conical and straight and the female gonopore lies on the ventral side at about one third of the entire length from the anterior end.

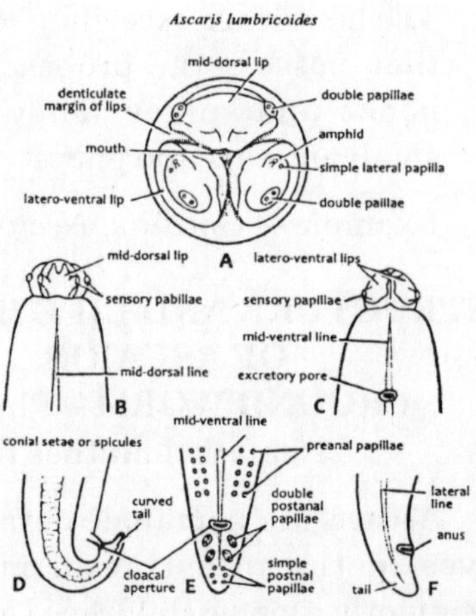

Ascaris lumbricoides. A - En-face view of lips and mouth. B - Anterior end in dorsal view. C - Anterior end in ventral view. D - Posterior end of male in side view. E - Posterior end of male in ventral view. F - Posterior end of female in side view.

Internal Structures

1. The body wall is thin consisting of the non-cellular, transparent cuticle secreted by the underlying epidermis. The epidermis is a thin protoplasmic layer containing nuclei without cell walls and hence is a syncytium. The muscular layer consists of only the longitudinal fibres.
2. Between the muscular layer and the intestine is the body cavity which is a pseudocoel as it is not derived from the splitting of the mesoderm unlike that of higher animals.
3. Digestive tract is a narrow straight tube divisible into (a) a club-shaped foregut or oesophagus, about 10-15 mm. long and lined with cuticle, (b) a voluminous midgut or intestine of endodermal in origin, and (c) a short hindgut or rectum discharging through the anus.
4. Circulatory or respiratory organs are absent. Respiration is anaerobic.
5. Excretory system includes two lateral longitudinal canals which open by a combined excretory pore behind the mouth.
6. Nervous system consists of a nerve ring, surrounding the oesophagus from which run six nerves forwards and six nerves backwards.
7. Sensory organs are the sensory papillae occurring on the lips and on the body surface.
8. Sexes separate. Reproductive system in the male is a single coiled tube running in the body cavity back and forth. The tube is differentiated into testis, vas deferens, seminal vesicle and ejaculatory duct opening into the cloaca. In the female, the ovaries occur in pair in the form of a slender coiled tubes. The tubes run upwards forming oviducts and uteri and then unite near the middle of the body as the muscular vagina that opens to the outside at the female gonopore or vulva.

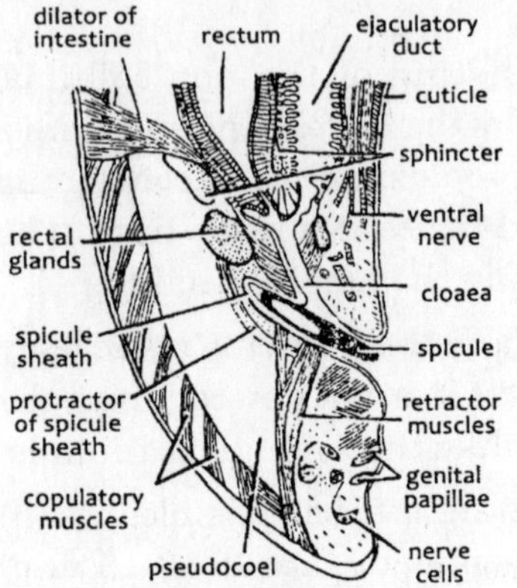

Ascaris. L.S. of the tail end of the male showing cloaca and spicules.

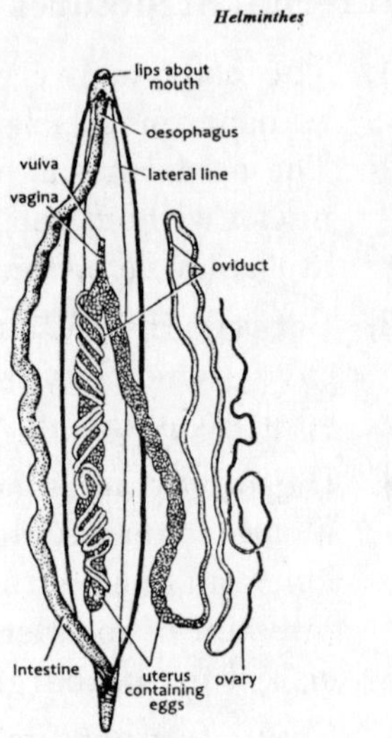

Ascaris. Female reproductive system

Life History

Ascaris is monogenetic. It completes its life cycle in one host, living only in man.

Copulation and fertilization: Copulation or sexual union takes places in the host's intestine. The sperms introduced into the vagina fertilize the ova within the oviducts. The zygote formed by the union of the sperm and the ovum become enclosed in a three layered chitinoid shell.

Egg laying: Each female Ascaris lays about 200,000 eggs per day. They are laid in the host's intestine, and leave with the faeces. The eggs survive in the soil for several months. Development is direct. Under favourable conditions of temperature and moisture, the

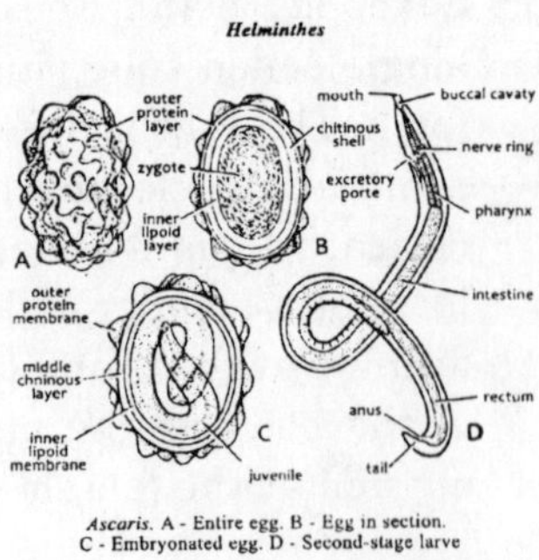

Ascaris. A - Entire egg. B - Egg in section. C - Embryonated egg. D - Second-stage larve

embryo develops inside the egg in about 8 days. The embryo is known as first stage juvenile. It continues its growth, and after about 10 days, it moults and becomes the 2nd stage infective juvenile.

Infection of host and larval wanderings: If the embryonated egg (containing infective embryo, juvenile) is swallowed by the human subject, with food or water, it reaches the intestine. There the infective juvenile emerges through a rent in the egg shell. The newly hatched larva is about 0.2 to 0.3 mm. in length. It burrows its way through the wall of the small intestine and to reach the lymphatics or venules. Blood carries it to the liver through portal circulation. In the liver, it lives for about 3-4 days. From there it enters the right auricle and then through the pulmonary artery it reaches the lungs. Here it undergoes two successive moults. As a result of moulting, the larva grows to a length of 1-2 mm. and its reproductive organs are developing. The juvenile leaves the blood stream, attacks the lung alveoli and trachea and passes into pharynx. Here it initiates the pharynged wall of the host and causes coughing. Consequently, the juvenile is swallowed and thus finally reaches the intestine. The entire tour is completed in 10-15 days. In the upper part of the intestine, it undergoes the fourth and the last moult in between 25th and 29th day. The young continues to grow and attains sexual maturity in about 6-10 weeks.

Ascaris lumbricoides

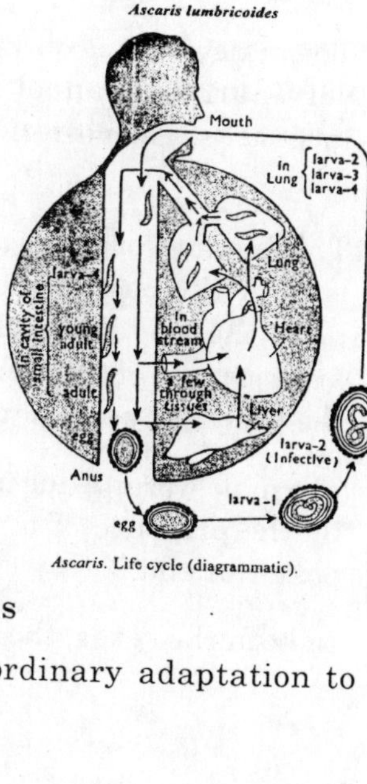

Ascaris. Life cycle (diagrammatic).

Parasitic adaptation of Ascaris

Ascaris exhibits few of the ordinary adaptation to a

parasitic mode of life. The adaptation include: (i) secretion of an irritating toxin; (ii) total absence of cilia; (iii) presence of thick and resistable cuticle; (iv) reduced nervous system; (v) absence of well organized sensory structures; (vi) production of innumerable number of eggs for continuity of the race.

STRUCTURE AND LIFE HISTORY OF *Trichinella Spiralis* (TRICHINA WORM)

Trichinella spiralis is placed in the Phylum Nemathelminthes, Class Nematoda. It is a nematode parasite found in adult stages in the intestine of man, pig and rat. It is common in Europe and America but it is absent in the tropics.

1. The trichina worm has a narrow hair like anterior end and a cylindrical broad posterior end.

2. The female worm is much larger than the male, and is from 3-4 mm. long, whitish in colour, slender and tapering from the middle of the body to the anterior end.

3. The male worm is only 1.5 mm. long. Its oesophagus is a very narrow channel running through the column of large cells. The posterior end of the body is occupied with testis.

4. The female worm has a single ovary and a coiled uterine tube at the posterior end of the body. The vulva opens near the middle of the oesophageal region, and the anterior part of the uterus is crowded with embryos, while the posterior part contains developing eggs.

5. The male worm, aside from the minute size is characterised by the presence of a pair of conical appendages at the posterior end.

6. In both the sexes, the anus or cloaca is terminal.

Life History

1. Copulation takes place within the intestine of the host. The males die soon after breeding and the impregnated females whose uterus becomes packed with developing eggs, and embryos, bore through the crypts of Lieberkuhn into the lymphatic spaces where they release their young, probably about 1500 larvae per female and then perish.

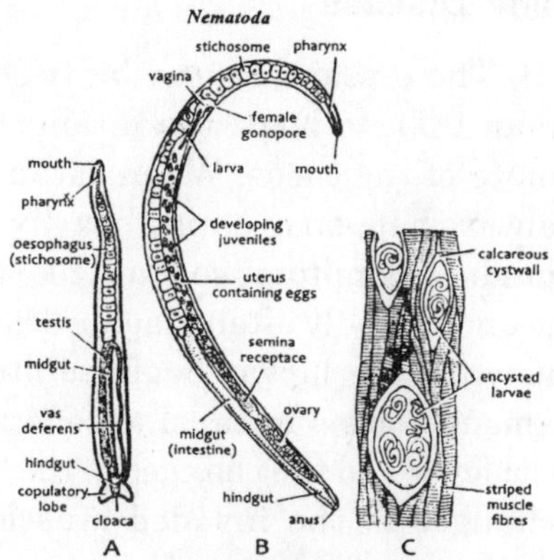

Trichinella spiralis. A - Adult male. B - Adult female. C - Section of striped muscles of pig showing encysted juveniles.

2. The juveniles, 0.1 mm. long, migrate by way of the lymph and blood vessels throughout the body, eventually entering every organ and tissue but maturing to the infective stage only in the voluntary musculature.

3. They enter the muscle cells and grow to a length 1 mm., being the 4th stage juvenile showing some sexual differentiation.

4. Now (at this stage) they coil up into a quiescent condition and the host animal forms a cyst around them that eventually calcifies.

5. Unless the host containing such encapsulated juvenile is eaten by a suitable host, the young worms degenerate and die in no time.

6. If, on the other hand, the dead body of a rat, a dead rat whose muscles were infected with encysted larvae, be eaten by another animal such as pig, or man eating

infected raw pork, the cysts are dissolved by the digestive juices, the worms escape from the cysts, come into the intestine, develop reproductive organs, copulate and the larvae migrate into host's muscles.

The Disease

The disease caused by trichina worms is more fatal to man than to hogs, the fatality sometimes rising to 30% or more of the cases. When eaten in large numbers, as would always happen in eating heavily infected raw pork, the worms produce symptoms so much like typhoid fever that the disease is undoubtedly often diagnosed as such. The really fatal stage is when the larvae begin to migrate. Trichinella is unique among worms in causing constant fever. It is probable that the fever and the changes in the blood corpuscles and chemical changes in the invaded muscles is due both to poisonous substances given off by the worms and to poisonous substances resulting from destroyed muscle tissue as shown by Flury and Groll.

4

PHYLUM ANNELIDA

Characters and Classification

The phylum Annelida (Latin, *annulus* = little ring) which includes Leech, Nereis, Earthworm, etc. is distinguished by the presence of the following characters:

1. The body is bilaterally symmetrical with three germ layers (ectoderm, endoderm and mesoderm).
2. The elongated body shows external and internal segmentation. (Segmentation or metamerism is a fundamental plan of body architecture in which the animal is composed of many essentially similar and ring-like segments or metameres).
3. In each segment of the body, there are minute, rod like chitinous appendages, called setae for locomotion. Some annelida may have fleshy tentacles on the head or, parapodia on lateral parts of the body.
4. Usually the head consists of a pre-oral "prostomium" and a post-oral peristomium.
5. The body is covered by a moist cuticle secreted by the underlying epidermis.

6. The body wall consists of both circular and longitudinal muscles.
7. The body cavity is a true coelome divided by septa (partioning wall) and communicating with the exterior by minute pores. The coelome is secondary in some Annelids.
8. The alimentary canal is straight, tubular and complete and divisible into distinct regions. The mouth at the anteriol and anus, terminal at the posterior end.
9. The circulàtory system is a 'closed type'. It essentially consists of longitudinal blood vessels with their lateral branches in each segment. The blood plasma contains dissolved haemoglobin and free amoebocytes.
10. Respiration is carried out by the moist skin; in some annelids gills are present for respiration.
11. Excretory organs are paired tubules arranged metamerically and are called the nephridia or segmental organs. Typically one pair of such nephridia occur per body segment, each collecting wastes from the coelome and general blood stream and removing to the exterior directly; nephridia may also act as genital ducts.
12. Nervous system typically consists of a pair of cerebral ganglia connected by a pair of connectives around the oesophagus to a solid (double) mid ventral nerve cord.
13. Sexes are separate in few cases; some are hermaphrodites; asexual reproduction and regeneration are common in majority of the Annelida. Development is direct or indirect; in the indirect development there is a larval stage called Trochophore.

Classification

The phylum Annelida has been divided as follows :

Class I. Chaetopoda

Animals included in this class are characterised by the presence of the following features :

1. The segments of the body are numerous and composed of almost similar segments. Setae are present in skin sacs or on parapodia.
2. The coelome is large and divided by intersegmental septa and not in communication with blood vascular system.
3. There is a well developed blood vascular system consisting of a dorsal or collecting vessel and a ventral or distributing vessel, and lateral hearts.
4. The nervous system consists of a double chain of segmental nerve ganglia connecting the paired, dorsal cerebral ganglia and extending ventrally throughout the length of the body.
5. The excretory organs are paired nephridia, a pair in each segment.
6. Sexes are separate or united. The gonads are formed from the coelomic epithelium.
7. Development is direct or indirect. In the indirect development a trochophore larva is formed.
8. They are found in freshwater, salt and blackish water ; some are terrestrial.

Order 1. Polychaeta

1. Usually a definite head with eyes, tentacles, cirri and palps; paropodia highly developed.
2. A clitellum is never formed.
3. Somites numerous, with well developed fleshy lateral outgrowths.

4. Sexes distinct, ovaries and testes of similar character.
5. Metamorphic development, and typically a trochophore larva.
6. Nearly all marine.

Sub order 1. *Errantia*

(i) A well developed head is present with a distinct prostomium.

(ii) Parapodia well developed with acicula.

(iii) Protrusible pharynx usually with jaws and teeth.

(iv) They are predatory and pelagic, burrowing and tube-dwelling worms.

Example – *Nereis, Aphrodite, Myzostoma,* etc.

Sub order 2. *Sedentaria*

(i) Head small and undeveloped.

(ii) Body distinguishable into regions.

(iii) No protrusible pharynx, or jaws or teeth.

(iv) Branchiae when present, are confined to the anterior region.

(v) They live in burrows or tubes.

Example – *Chaetopterus, Arenicola, Terebella,* etc.

Order 3. *Oligochaeta*

(i) No parapodia.

(ii) Head not distinct.

(iii) A clitellum is usually present.

(iv) Reproductive system complicated and with sexes united.

(v) No metamorphosis.

They include terrestrial forms like Earthworms (Pheretima, Lumbricus, Megascolids, and aquatic forms like Tubifex, Chaetogaster, etc.)

Class II Hirudinea

1. Body composed of small number of segments, marked externally by secondary ring or anuli ; no setae, parapodia or tentacles.
2. There are two suckers one at each end of the body, the anterior being formed by the mouth ; anus dorsal immediately in front of the posterior sucker.
3. Coelom is more or less obliterated by connective tissue and muscles and broken into sinuses filled with blood.
4. Excretory organs include segmentally arranged paired nephridea.
5. Hermaphrodite with numerous testes and one pair of ovaries.
6. Development is direct without a larva.

Order 1. Acanthobdellida

1. No anterior sucker, proboscis or jaws.
2. Setae present in the five anterior segments.
3. Coelom spacious and segmented.
4. Parasitic on fishes.

   Example – *Acanthobdella.*

Order 2. Rhynchobdellida

1. Pharynx protrusible as the proboscis, no jaws.
2. With or without anterior sucker.
3. Blood plasma colourless.

4. Strictly aquatic.

Example – *Glossiphonia, Branchellion,* etc.

Order 3. Gnathobdellida

1. Anterior sucker surrounding mouth suctorial, no proboscis.
2. Pharynx with three jaws.
3. Blood is red.
4. Aquatic or terrestrial.

Example – *Hirudo medicinalis* (European leech), *Hirudinaria* (large cattle leech of India).

Order 4. Herpobdellida

1. A definite anterior sucker is present ; posterior sucker is small.
2. Mouth is overhung by the projecting lip.
3. Jaws absent.

Example – *Nephelis, Orbdella,* etc.

Class III Archianellida

1. Body is cylindrical, external segmentation indistinct, internal segmentation distinct, marked by septa.
2. Head is with a pair of short tentacles and a pair of ciliated pits, no parapodia, no setae.
3. Pairs of simple types of nephridia are present.
4. Nervous system epidermal.
5. Vascular system consists of a dorsal longitudinal tube and a ventral tube.
6. Sexes separate, in a few united; gonads in posterior segments, numerous with no special reproductive ducts ;

a Trochophore larva in development.

Example – *Polygordius, Protodrilus*, etc.

Class IV Gephyrea

1. Body unsegmented, without parapodia or suckers; anterior end is drawn out into a proboscis or involuted forming introvert carrying lobed and plated folds termed the tentacular fold.
2. Alimentary canal is complete with mouth and anus.
3. Coelom spacious but not divided by septa.
4. Nephridia are present.
5. Blood vascular system closed.
6. Sexes separate; gonads on the body cavity; a trochophore larva present.

Order 1. Echiurida

1. Body consists of a sausage – shaped cylindrical trunk and an anterior poboscis. The poboscis is a cephalic lobe.
2. The trunk may be smooth or ornamented with papillae that are irregularly arranged in rings round the body.
3. Porapodia are absent.
4. Alimentary canal is greatly coiled; mouth is at the base of the proboscis; into the rectum there opens a pair of anal vesicles.
5. A pair of nephridia are present.
6. Blood vascular system is closed.
7. Nervous system resembles other annelids.
8. Sexes are separate; development is with a Trochophore larva.

9. They are either marine or fresh water.

Example – *Echiurus, Bonellia.*

Order 2. Sipunculida

1. Body is large, cylindrical in shape; anterior body part or introvert can be inverted, and the integument is folded so as to have the appearance of a circle of tentacles.
2. The surface of the body is covered by a chitinoid cuticle.
3. The gut is much coiled and the anus is anterior and dorsal in position.
4. Coelom is large and undivided by septa.
5. A single pair of nephridia which acts as reproductive ducts, is present.
6. Sexes are separate and larva is a trochophore.

Example – *Sipunculus, Phascolosoma,* etc.

Order 3. Priapulida

1. Body is cylindrical with superficial annulations in the posterior half of the body.
2. At the posterior end is a hollow process provided with papillae called branchiae for supposed respiratory function.
3. No tentacular folds, but there is introvert.
4. Digestive tract is straight.
5. No vascular system and nephridia.
6. Sexes are separate.

Example – *Priapulus, Halicruptus,* etc.

STRUCTURE AND LIFE HISTORY OF NEREIS

Nereis (Rag worm or Clam worm or Sandworm) is

cosmopolitan in distribution. Various species of Nereis occur abundantly among sea-weeds and under stones between tide marks. They vary considerably in colour though they invariably have richly tinted iridescent hues. The common species of Nereis are – N. dumerilli, N. pelagica, N. virens, N. brandti,N. limincola, etc.

External features of Nereis

The body of Nereis is long, narrow and slender, and tapering posteriorly, reaching a length of 7-8 inches. The upper surface of it is slightly convex and the lower surface is flat. The anterior end bears a distinct head, while the posterior extremity is conveniently called the "tail". The portion of the body lying in between the head and the tail is marked by a series of more or less equal segments or metameres, which are about 80 in number altogether.

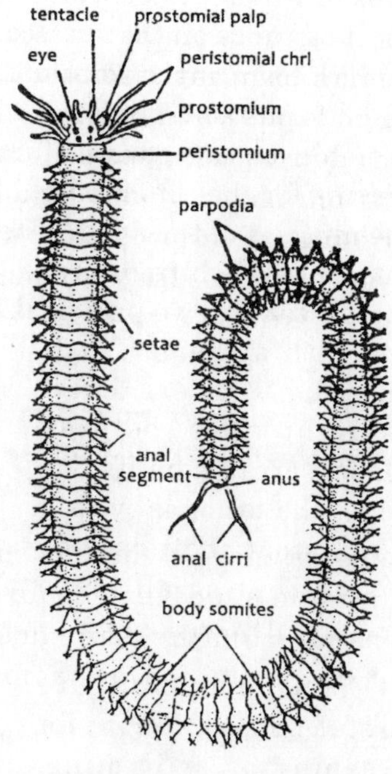

Neanthes. External features.

Each segment of the body bears a pair of movable muscular processes called the parapodia. A parapodium is a lateral flap-like projection of the body wall bearing bristles or setae. It is biramous, consisting of a basal portion and two lobes, of which one is dorsal (or upper part) and known as notopodium; and the other ventral (or lower part) called the neuropodium. Both the notopodium and neuropodium are bilobed and each bears a bundle of setae embedded in a sac, the setigerous sac formed by an inpushing of the epidermis.

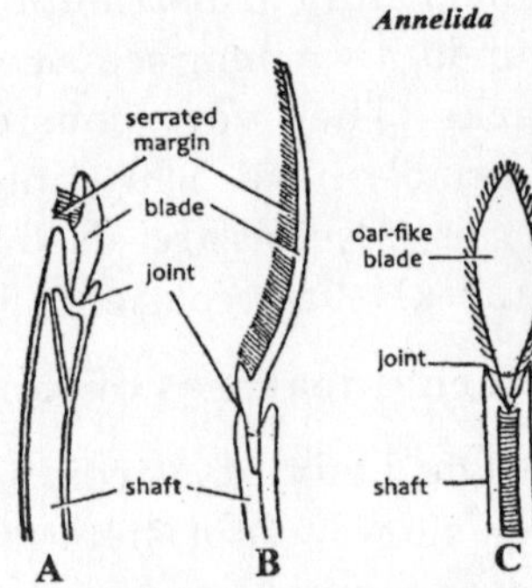

Nereis. Kinds of setae.
A - Typical. B - Long-bladed.
C - Oar-shaped seta of *Heteronereis*.

In each bundle there is, in addition to the ordinary setae, a stouter, straight dark coloured seta, the pointed apex of which projects only a short distance on the surface: this is known as the aciculum, and it forms a sort of internal skeleton to the parapodium serving for the attachment of the muscles of the setae. The ordinary setae are exceedingly fine, but stiff chitinous rods. Two principal kinds of setae are recognizable. Two principal kinds of setae are recognizable: one with a slender shaft and long, slender, almost straight and pointed blade, serrated on its one edge, the other is with a large, stout shaft bearing a stout but small blade and slightly hooked.

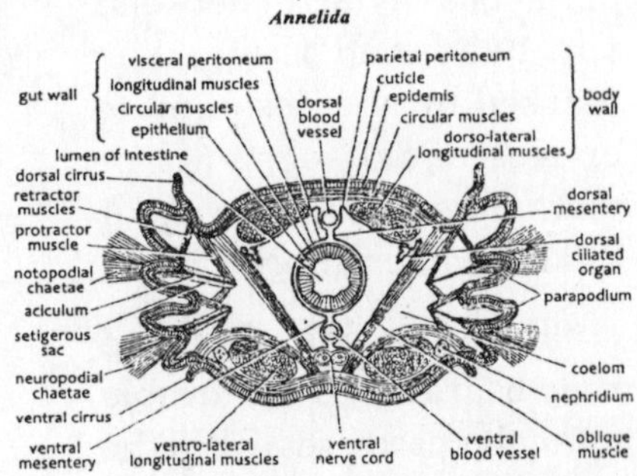

Neanthes. Transverse section of a segment with parapodia.

The parapodia are useful in locomotion, for creeping or swimming; they are highly vascular and subserve also the function of respiration.

The head is divisible into two parts: (i) a roughly triangular anterior lobe, called the prostomium which forms the anteriormost part of the body, (ii) a ring like portion, called the peristomium lying just behind the prostomium.

Projecting forwards from the anterior border of the prostomium is a short pair of tentacles, and from the ventral side of the prostomium a pair of short and stout 2-jointed palps. Both of these types of organs are probably tactile in function.

COELOM IN ANNELIDA

Coelom or secondary body cavity is a wide space extending from one end to the other of the body so as to surround the alimentary canal. The coelom also encloses other internal organs, such as the nephridia and the reproductive organs. The annelids have a true coelom formed by the splitting of the mesoderm during development. It is bounded on all sides by a definite coelomic epithelium. Reproductive organs arise from its walls, and the excretory organs open into it. The coelom is filled with a colourless coelomic fluid which contains amoeboid cells. The coelomic fluid serves to destroy the harmful micro-organisms which may adhere to the soft body. This fluid continuously oozes out of the body through the dorsal pores and keeps the skin moist to help in respiration.

A large portion of the coelom forms a spacious perivisceral cavity or sphlanchnocoel between the digestive organs and the body wall, but certain portions cut off from it to form gonocoel and nephrocoel.

Developmentally, the coelom in Annelids, appears in the late gastrula stages as splits in the mesodermal somites. Typically one pair of sacs, the right and left coelomic vesicles is found in each segment of the body of the embryo. Since the mesodermal somites occur in pairs, the two coelomic cavities extend upwards and downwards and their walls unite in the mid-line above and below the gut forming the dorsal and ventral mesenteries. The ventral mesentery usually disappears in the adult so that the right and left coelomic spaces become confluent below the gut. Each coelomic space has a layer called the Somatopleure which is closely attached to the body wall, and a layer known as splanchnopleure applied to the alimentary canal.

When the coelomic cavity increases in size, the walls of the mesodermal somites extend anteriorly and posteriorly and meet with similar structures nearby. In this way, horizontal septa are formed between the coelomic cavities of the body segments. The septa keep the alimentary canal in position, and they also divide the body cavity of the entire animal into a number of compartments, corresponding to the external segmentation of the body. Each septum consists of two layers of peritoneum enclosing muscle fibres. The intersegmental septa and the mesenteries are generally perforated and incomplete so that the coelomic cavities of each segment intercommunicate freely with one another.

Coelom in Oligochaeta

In Oligochaeta coelom is a wide chamber enclosing internal organs, such as the nephridia and the reproductive organs. The alimentary canal runs through the coelom. Typically, in Earthworm, the coelom is transversely divided by septa into a number of compartments. Each septum runs in a transverse direction from the body wall to the wall of the alimentary canal. The septa are intersegmental in position. But anteriorly beyond the 4th segment, the septa are absent. The septa which occur in the 6th and 9th segments are thick and muscular. Those in the middle and posterior regions of the body following the 15th segments are perforated

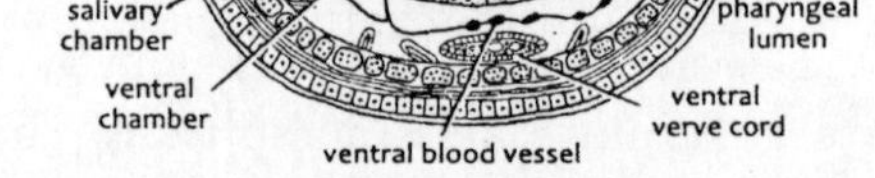

Pheretima. T.S. body through the pharyngeal region.

through which the coelomic compartments are intercommunicated. These perforations are provided with rings of muscles which can close the apertures by contraction.

Coelom in Polychoeta

The coelom is a spacious perivisceral cavity divided into a linear series of compartments by cross partitions or septa as in the Earthworm. Each septum consists of a double layer of coelomic epithelium, connective tissue and muscles, and in position each corresponds to an external groove. The septa are provided with numerous apertures so that the compartments are not completely isolated. Furthermore, the coelom is partially subdivided by various muscles which run an oblique course from the dorsal to the ventral body wall. In burrowing and tubicolous worms, the septa are incompletely developed, or may even be absent. Owing to the reduction of the septa, the coelomic fluid flows freely between the segments, and this facilitates in the enlargement of the anterior end of the body during the course of burrowing.

Coelom in Archiannelida

Coelom is spacious, undivided but traversed by many strands of connective tissue and muscle attaching to the gut.

Coelom in Hirudinea

In Hirudinea the coelom is greatly reduced as its space is taken up by a spongy connective tissue (botryoidal tissue) that fills the space between the body wall and the alimentary canal, and this replaced the perivisceral cavity of the coelom. Among the Hirudinea, it is only in Acanthobdella that a spacious coelom divided by septa occurs as in Oligochaete worms. In all other forms, the remnants of the coelom are represented as sinuses and channels connected by a complicated system of inter communicating spaces. In

Hirudinaria there are four haemocoelomic longitudinal canals, two lateral, one dorsal and one ventral. Through the coelomic canals circulates the coelomic fluid. All the canals are thin walled. Further, in Hirudinaria, the true coelom containing fluid without the respiratory pigment is restricted to the testis-sac and ovisacs.

Hirudinaria granulosa

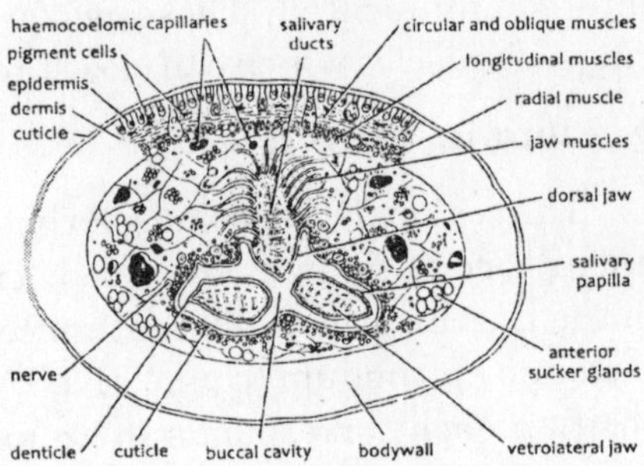

Hirudinanaria T.S. body through buccal cavity.

In *Rhynchobdella,* the dorsal sinus encloses the dorsal blood vessel and the ventral sinus encloses the ventral blood vessel and ventral nerve cord. The sinuses and channels are filled with coelomic fluid which resembles blood in having free amoeboid cells and dissolved haemoglobin. This type of coelomic fluid is described as haemocoelomic fluid.

In *Sipunculida,* coelom is spacious, undivided but traversed b many strands of connective tissue and muscle attaching to the gut. It is filled with coelomic fluid in which float several types of corpuscles.

In *F apulida,* a cavity exists between the body wall and the digestive tract but the true nature of this has not yet been found out. It is lined by structureless membranes having r. nuclei. The space is filled with a fluid into which float large number of round cells.

TROCHOPHORE LARVA STRUCTURE AND AFFINITIES

The Trochophore larva is a type of larva common to Polychoeta, and Archiannelida among Annelida and the phylum Mollusca. Polychaetes undergo spiral determinate cleavage, closely resembling that of Molluscs and Flatworms. After gastrulation, a Trochophore larva develops, which corresponds to the Trochophore stage of Molluscs. After hatching out from the egg, the Trochophore settles to the bottom to metamorphose into the adult form.

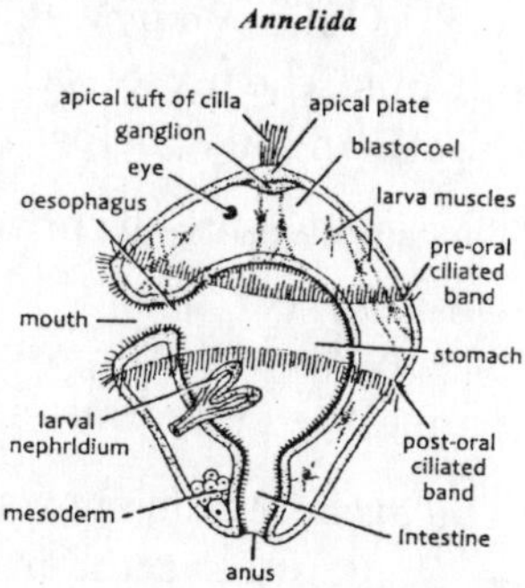

A typical trochophore larva.

Typically, a Trochophore larva is a minute, unsegmented ciliated and almost pear-shaped pelagic creature with a protruding equator and the anterior end (head end) broader than the posterior end (anal end). The larva shows a distinct bilateral symmetry. Its external surface consists of a one-layered epithelium which is thickened only at the following regions.

(a) anteriorly to form the apical plate which bears a ciliated tuft-apical tuft-both of which are sensory in function;

(b) at the elongated region where it is ciliated to form the prototroch situated anterior to the mouth, often a second metatrich behind the mouth may be present ; a telotroch surrounding the posterior extremity may also occur;

(c) a ventral plate formed of a thickening of the epithelium in the median line ventrally which form the nerve cord.

There is a well developed alimentary canal. The mouth lies in the middle of the body, and it leads into a short oesophagus which opens behind into the wide stomach. Stomach is followed by a short intestine which opens outside by the anus. The prototroch serves to direct the food towards the mouth and also helps the animal in swimming.

The whole epithelium of the post-oral region is formed of the cells derived from the 1st somatoblast of the gastrula, and it goes to form the whole ventral and lateral regions of the trunk, the setigerous sac and the nerve cord.

The archenteron shows a distinct curve and is formed of three elements – ectodermal invagination anteriorly and posteriorly called stomoderm and proctoderm respectively and a midgut of ectodermal origin.

Between the gut and the body wall, there is a wide spacious cavity – the remnants of the blastocoel, traversed by strands of muscular tissue formed of larval mesoderm or mesenchyme derived from cells which have migrated from the ectodermal layer. The true mesoderm which originates from the second somatoblast or mesoblast of the blastula, is still in the process of making.

Larval nephridia are a pair of long tubular, excretory organs originated from a cell which has migrated inwards from the ectodermal layer – one on either side of the gut. Each nephridium is formed of two hollow cells placed end to end one of which is provided with a flame of cilia while the other opens to the exterior near the anus. An otocyst lies near the nephridia.

Metamorphosis of Trochophore

The trochophore is a pelagic creature drifting about in the sea. While swimming it undergoes metamorphosis to become the adult. The change to adult form probably takes place in the most primitive fashion in the little Archiannelid worm, Polygordius. The development of the Trochophore into the adult worm begins with the elongation of the lower region of the Trochophore. The elongated region becomes constricted into segments which soon develop bristles. The ciliated bands disappear and the part of the larva between the apical tuft and a point somewhat below the mouth develop into the head of the young animal. When it is present, the line of the metatroch lies near the posterior margin of the future head. The telotroch and anus develop into the final somite of the body, the pygidium. All of the trunk arises from the short, tapered region between. During metamorphosis, this part of the larva grows very rapidly; it may be termed the growth zone. This growth zone grows steadily and rapidly. As it lengthens, the anterior part is cut off as the first post cephalic somite. It continues to grow, and the second somite forms behind the first. New tissue from the growth zone undergoes some maturation process. Setal sacs and setae differentiate in the new somites. A girdle of cilia around each somite often persists throughout the planktonic phase of larval life. As the somites form, paired spaces appear in the mesodermal rudiment of the somite, eventually enlarging to form the right and left coelomic compartments of the segment. All of the ectodermal covering of the trunk arises from the second quartet of micromeres. Ventral nerve cord ganglia arise from the ventral plate of ectodermal cells. The brain develops from the apical sensory plate. The part of the larva around the plate develops into the prostomium, and grows tentacles and palps on each side.

In polygordius, the prostomial region is separated from the rest of the body by a large part of the Trochophore. As growth continues, this part of the Trochophore collapses, partly as a result of the contraction of larval muscle strands, bringing the prostomium back to the developing trunk. The prostomioum develops from the collapsed part of the Trochophore, which contains the mouth and through which the connectives to the suprapharyngeal ganglion pass. In many Polychaetes the suprapharyngeal ganglion lies in the peristomium. In this case the adult peristomium is formed by the union of the first trunk somite and the peristomial region of the Trochophore, as in Nereis. As development continues, the larva settles to the bottom and abandons its pelagic life, and becomes the adult worm.

Affinities of Trochophore larva

Trochophore larva has great phylogenetic significance. A trochophore larva or a stage very similar to it with modifications occurs in the development of animals belonging to diverse groups such as Nemertinea, Polyclad Turbellaira, Brachiophoda, Bryozoa, Annelida, Mollusca, and Echinodermata. Among these groups the early stages of development are almost identical, while the trochophore stages are similar in many respects. Occurrence of such a common larval pattern is suggestive of a relationship between different phyla or groups of animals. However, marked differences arise beyond the trochophore stage so that the adults become very different.

The Recapitulation theory states that these groups of animals were descended from an ancestor similar to a trochophore. Since, however, new characters are added even at the larval stage a conclusion is reached that these groups of animals were descended from an ancestor in which a trochophore larva existed.

According to the Trochophore Theory as propounded by Hatschek (1878) and other embryologists, most of the present day bilateral phyla descended from a common hypothetical ancestor, the Trochozoon which evolved from the Ctenophora.

Affinities of Trochophore larva

I. With Ctenophora

Similarities

1. Body form is pear-shaped in both Ctenophora and Trochophore.

2. The aboral sense organ or statocyst of Ctenophora is comparable to the apical sensory plate of Trochophore.
3. Both Ctenophora and Trochophore possess more or less similar types of sub ectodermal radiating nerves.

Dissimilarities

1. On closer examination, the body organization of Ctenophora exhibits dissimilarities with Trochophore
2. Trochophore possesses anus, but it is absent in Ctenophora.
3. The cleavage pattern of Ctenophora differs from that of Trochophore.

II. Affinities with Muller's (Polyclad) larva of Turbellaria

Similarities

1. Ciliated bands are more or less similarly arranged in both.
2. Eye spots are present at the aboral end in both.
3. There are similarities in the development stage of both.

Dissimilarities

1. There is no anus in the Muller's Polyclad larva but it is present in Trochphore larva.
2. The enteron has only one opening in Muller's Polyclad larva.
3. Muller's Polyclad larva possesses a tuft of cilia at the caudal end.
4. Embryonic differentiation of mesoderm is not similar in both.

III. Affinities with Pilidium (Nemertini) larva

1. Body form is more or less similar.

2. In Pilidium there is a ciliated ring between the aboral and oral ends, and this resembles the prototroch of Trochophore.
3. The formation of coelom is schizocoelic mode in both.
4. Disposition and distribution of nerve ring are similar in both the cases.

IV. Affinities with Rotifera

1. Presence of ciliated girdle in both.
2. Nephridia present in both.
3. Distributions of nervous system and sense organs are similar in both.

V. Affinities with Veliger larva

The Veliger larva of Molluscs has certain similarities with that of Trochophore larva in having more or less pear-shape, a tuft of long cilia on the top as well as a ring of cilia (prototroch) and a girdle of cilia in the middle region.

Poctella (Gastropod) has typical trochophore stage. A trochophore stage similar to that of Patella also occurs in other Gastropods but in a modified form.

The marked similarities in larval forms, but not exhibited by their adults, largely because the differentiations which characterize the adults have not yet appeared, had led to the conception that the larval forms like trochophore, veliger (in Mollusca), Pilidium (in Nemertinea) or Pluteus (in Echinodermator), Tornaria (in Balanoglossus) represent the adult form of the ancestors as the theory of Recapiluation tried to make it out. It has been accepted that these larval forms represent adaptive modifications to a pelagic habit fulfilling the need of dispersal of the species. Thus, as a corollary to this, the better is the larva suited for its mode of life, the greater will be the difference between it and the adult and the metamorphosis will be more striking. Nevertheless the similarities indicate the inherited qualities which further furnish as a sort of evidence of affinity between groups having a similar type of larval form.

BLOOD VASCULAR SYSTEM IN ANNELIDA

The blood vascular system is that system that performs the carriage of blood along with carbon dioxide, gives off carbon dioxide as the metabolic wastes.

The blood vascular system has been developed for the first time in the Phylum Annelida among the whole animal kingdom. Its development is obviously, due to the complication of the body structure and the associated requirement of the carriage of food from the gut and respiratory gases from the skin to all tissues of the body and also for the removal of the waste products of metabolism from the tissues of the excretory organs. All these requirements are fulfilled by the annelid blood vascular system.

Blood

Blood consists of fluid part, plasma, and corpuscles. Usually it is coloured but may be colourless also. When coloured, which is due to pigments confined not in corpuscles but dissolved in plasma, it may be red (due to haemoglobin), or green or yellow or pink.

Circulatory apparatus

(i) In Annelida, there are two radically different blood vascular apparatus – one is undifferentiated from the body cavity (haemocoel) and the other is entirely separated from the body cavity.

(ii) Two longitudinal blood vessels – one dorsal above the gut and one ventral below the gut with collateral connections and joined at either end by capillary networks are present.

(iii) Additional longitudinal vessels – sub-neural, lateral, neural are also present.

(iv) There are main vessels giving off branches to supply different organs.

(v) There is a contractile or pulsatile part which may be a part or whole of dorsal or lateral loops joining dorsal and ventral vessels or intestinal sinus (if present) or other structures.

(vi) Valves are also present in the different parts of the system.

In the Annelids blood vessels are related to each other in the following fashion : –

Blood vascular system (circulatory apparatus) in different classes of Annelids

I Class Chaetopoda

Typically there are –

(i) two most important vessels – dorsal and ventral running from one end of the body to the other. In the dorsal longitudinal vessel blood runs from behind forward, and in the ventral longitudinal vessel blood runs from forwards to backwards.

(ii) There is a contractile or pulsatile part of the dorsal vessel, and this serves to force the blood through circulation.

(iii) Two pairs of loop-like transverse vessels in each segment connect the dorsal vessel with the ventral one. Both the vessels give off segmentally arranged vessels to the body wall, gut, parapodia, etc.

(iv) Additional longitudinal vessels – subneural, lateral, neural, etc., may also be present.

In the order Polychaeta, besides the typical condition, the following variations are met with.

(i) In the family Terebellidae, the dorsal vessel forms a pulsatile heart anteriorly. Posteriorly it divides into two branches which encircle the gut and join below the intestine into a sub-intestinal vessel. The latter runs for a short distance and divides into two and meet above the intestine to form a dorsal vessel running to the hind end of the body.

(ii) In Cerratulidae, at posterior region of the body where branchiae are absent, a separate dorsal vessel and an enteric vascular plexus are both wanting. They are here replaced by an intestinal blood sinus which is continued anteriorly into a strong pulsating dorsal vessel (heart) which runs through the whole branchial region and gives off vessels to gills directly or indirectly.

(iii) In Serpulidae, a dorsal vessel is entirely absent. Instead, there is an intestinal sinus running throughout the whole length of the body. A pair of lateral and ventral vessel are present.

(iv) In Capitellidae and Glyceridae, a blood vascular system separate from the body cavity is wanting. The coelomic fluid, haemolymph performs its function.

In the order Oligochaeta, in addition to a dorsal and a ventral trunk, the following vessels are found:

(i) a subneural vessel which runs under ventral chord;

(ii) two lateral vessels of ventral chord which accompanies the ventral chord throughout its whole length;

(iii) variable number of paired lateral vessels (5 pairs in Lumbricus), (6 pairs in Pheretima, etc.), the wider pouch-like contractile vascular loops – the so called hearts, connecting the dorsal vessel with the ventral one.

(iv) a peri-enteric sinus may be found.

II Class Hirudinea

(i) In Pontobdella, there are four vessels – a dorsal, a ventral and two laterals all of which are enclosed in sinuses.

(ii) In Claspine, there is a dorsal and a ventral vessel both contained in sinuses.

(iii) In Leech (Hirudo), there are two lateral vessels not enclosed in sinuses, having muscular walls, and a dorsal and a ventral sinus, which have an appearance of vessels, devoid of muscular walls.

(iv) All the various vessels are connected by anterior and posterior loops or only by the anastomoses of their terminal capillaries.

The dorsal and ventral vessels are developed in the manner of true vessels in all Hirudinea, except Gnathobdellida where these are coelomic in origin like the laterals.

Variations

In Nephelis and some land Leeches, there are, in connection with anastomoses between the lateral and ventral vessels, a ıpullae or blood vesicles in segmental arrangement, one on ea h side (Land leech) or two together (Nepheles).

In Branchellion, on each side, at the base of every third g ll, there is a blood sinus, widened into a vesicle.

III Class Echiurida

The dorsal vessel is short, and extending to the posterior end of the foregut only. It is joi ıed at both the ends by two loops with the ventral vessel, which, however, runs throughout the whole body.

IV Class Sipunculida

The blood vascular system is feebly developed. It consists of contractile dorsal and ventral vessel (the former is known as the heart), opening anteriorly into the circular sinus at the base of the tentacular fold.

V Class Archiannelida

The vascular system consists of a dorsal vessel in the dorsal mesentery and a ventral vessel in the ventral mesentery and of connective vessels. The blood may be red, green, yellow or colourless (*Ex* – Protodrilus).

Mechanism

The movement of the blood is effected in most instances by peristaltic contraction of the dorsal vessel or of a peri-intestinal sinus or plexus, or of the 'heart' given off by the latter anteriorly, which have the effect of driving the blood from behind forwards. In some instances, as in the Earthworms and some Sedentaria specially dilated lateral vessels are contractile, and by their pulsations bring about the circulation of the blood through the system of vessels. In dorsal blood vessel flow is from behind forwards while in ventral vessel it is from before backwards. The valves maintain one way flow of blood.

REPRODUCTION IN ANNELIDA

Reproductive organs in Annelida are the products of coelomic epithelium, thickened to form papillae. Occassionally in males, more often than females, the generative section of the coelom is cut off from the rest by special membranes, as in sperm reservoir of some Oligochaeta. The generative ducts, however, open into the coelom and have a tubular character, with a ciliated internal opening. Sometimes they are distinct from the nephridia, but as a rule, some or all of nephridia are used for the exit of the generative products, thus combining a renal and genital function. Thus, the coelom functions as the carrying apparatus for both generative and excretory products.

The reproductive function is carried on in different individuals of the phylum Annelida, in different ways. In general they may be classified into two different groups according to the modes of performance of the function: (1) Asexual and (2) Sexual. In asexual mode, sex is not involved while the other (sexual) is performed through sex. The sexual mode again may be classified according to the mode of occurrence of generative organs in different individuals into two viz., Unisexual and Bisexual. In the unisexual case, the individuals show a sexual dimorphism and in the bisexual case, the individuals are hermaphrodite.

I Asexual reproduction

This occurs in certain forms of Chaetopoda, Polychaeta and Oligochaeta. The peculiarities of these groups are, that they can regenerate their lost parts and can grow indefinitely throughout their life. This power of regeneration and indefinite growth are associated with the power of reproducing asexually.

Following are the examples in different groups: –

Polychaeta

1. Asexual reproduction by budding is common in many forms where the number of segments increases throughout life and the animal continually grows in length.

2. The power of regeneration in many of them helps in asexual reproduction. This power is very extensive, and they can reproduce any part of the body when cut into pieces, thus producing a number of individuals.

3. Asexual reproduction by simple transverse fission occurs in Salmacina, one of the Serpulids. A constriction appears near the posterior end, new cephalic branchae appear on one side and at last it is detached from the anterior part and becomes a distinct zooid.

4. Schizogony or metagenesis is another type of asexual reproduction found in some genus of Syllidae, where the sexual forms are produced asexually by sexless forms by a process of gammation.

   (a) The heterosyllid condition is first assumed and the worm divides into an anterior non-sexual and a posterior sexual portion. The posterior portion then separates out and acquires a new head, and thus forms either a male or a female worm. The anterior portion also acquires again the anal segments and in turn again forms the sexual worm separating out from the asexual part.

5. In Autolytes, the process is still more complicated. The formation of new segments begins before the two parts of the worm separate.

   (a) A zone of fission is formed and this zone may produce not only the head of the hinder zone but a number of segments which become marked out into a series of zooids in which genital organs eventually appear.

   (b) In this way a chain of sexual zooids, attached to the sexual "scolex" in front, is formed.

   (c) The zooids are all of the same sex, and the posterior of them is the oldest and most developed.

   (d) The male zooids differ very markedly from the female.

6. In Myrianida a very peculiar process takes place. A chain of sexual zooids attached to the anterior sexless zooid is formed. In this case there is no fission but gammation

takes place from a zone of fission, for non of the segments of the original worm, except the anal enter into the constitution of the first formed sexual zooids. The process starts with the formation, in the gemmiparous zone between the penultimate and anal segment, which give rise successively to the first formed and all the subsequent zooid of the chain. The origin of the male and female is the same as in Autolytus.

7. *Syllis ramosa* which lives in the canal system of an Hexactienellid sponge, possesses a peculiar property of budding and fission. It gives off lateral buds which send out branches that produce buds, some of which develop heads and generative organs so that an arborescent (tree like branches) colony of zooids is formed. The sexual branches separate from the colony and distribute the ova.

8. *Polymorphism in Nereis*

   It has been shown by *Claparade* that the worms undergo some remarkable changes during maturation to a form known as heteronerried condition. These changes chiefly affect the posterior end of the body in which the generative organs are combined. The changes are – (a) In the posterior part the parapodia become larger and acquire flattened, foliaceous outgrowths. (b) Old setae are thrown off and are replaced by new setae of flattened form and with a fan-like arrangement. (c) The eyes become enlarged. (d) Dorsal cirri are flattened. (e) Animal becomes free swimming. (f) Sexual dimorphism sets in.

In *Nereis dumerilli*, *Claparode* has shown that there are four distinct sexual forms differing from one another in size, form, and mode of life, etc.

(i) A sexual, dioecious Nereis form, distinguished by its small size.

(ii) A sexual hermaphrodite Nereis form.

(iii) A dioecious Heteronereis which is small and it swims on the surface.

(iv) A dioecious Heteronereis which is larger, and creeps to the bottom.

9. In some of the members of Syllidae, there is a distinct alternation of generations. The asexual worm developed from the ovum gives rise, by a process of proliferation and constriction, to sexual zooids, many of which may remain connected for a time in a string. After separation, the sexual zooids become developed into mature males or females. The sexual forms give rise to ova and sperms which after fertilization, gives rise to asexual worms.

Oligochaeta

In some Oligochaeta, asexual mode of reproduction is also found.

1. In Nais and Chaetogaster, the process of proliferating or budding new individuals is well developed. By the process of proliferation and constriction a chain of zooids results. These zooids remain connected together for a time. The sexual cells become fully developed after the separation of the zooids from one another.
2. Regeneration is also a common occurrence in some Oligochaets like the earthworms, where the body of the animal when cut into pieces can regenerate its lost parts and can form new earthworm.

II Sexual reproduction

The mode of sexual reproduction differs according to the occurrence of organs of sex in different Annelids. In the arrangement of the reproductive organs in Chaetopoda there is an essential difference between the two orders, the

Oligochaeta being hermaphrodite and Polychaeta with only a very few exception, unisexual. Usually the sexual organs, male or female, originate as proliferation of coelomic epithelium of the hinder region of the body cavity, ventrally during the breeding season. They are papillae like in structure and lie on both sides of the ventral vessel; they are usually in pairs, one or more, and may or may not be provided with definite ducts for conveying their products outside.

Archiannelida

In them sexes are separate. The sexual organs are developed from the coelomic epithelium by simple proliferation. The sex cells – ova and sperms are discharged into the body cavity. In Polygordius, the ova are discharged by rupturing the body wall and the sperms by nephridia. Fertilization is external.

Polychaeta

1. All the Polychaets with a few exceptions are unisexual.
2. Sexual organs – testes and ovaries occur in majority of the segments. The exact position of their occurrence in the segment is not always fixed, and in some cases, they are placed laterally at the base of the parapodia.
3. The final stage of development of sperm takes place after it has been detached from the testes, and while floating in the coelomic fluid the same also holds good for the ova in some cases.
4. The ova and sperms are discharged through the segmental organs. But in some forms, they escape through the temporary or permanent opening of the body wall.
5. Most of them are oviparous, where the fertilization is external with a few exception, while the animal is viviparous as in *Syllis vittata*.

Echiurida

1. All the members of this group are unisexual.
2. Sexual organs of this group are developed from the portion of the coelomic epithelium which overlies the ventral vessel.
3. The maturation of the sex cells takes place after they have been discharged into the body cavity and then escaped through anterior nephridia.
4. Extreme dimorphism is observed in some cases like Bonellia and Hamingia where the males are minute and they live into uterus or pharynx of larger female. Fertilization is external.

Sipunculida

1. Generally the sexes are separate.
2. Sex cells are discharged from the genital ridges in immature condition into the body cavity and maturation takes place there.
3. Nephridia act as gonaducts. Fertilization is external.

Bisexuality or Hermaphroditism

(a) Polychaeta

1. Only a few of the Polychaets are hermaphrodite, for example, Sabelliformia, Salmacina, Protula, etc. The sexual cells originate and come out of the body in the same manner as in the unisexuals.

(b) Oligochaeta

1. The Oligochaeta are all hermaphrodites. Reproductive organs are continued to a limited region of the body.

Male

1. There are either 2 pairs (Earthworm) or one pair of testis (in aquatic forms).

2. The testes are small and become reduced to mere vestiges in adult condition, to be broken up into sperm mother cells and in some way reaching vesicula seminalis to undergo further development into mature sperm.
3. The vesiculae seminalis are comparatively large sacs which vary in number and arrangement in different genera
4. In the same segment as the testes there are two or four ciliated funnels according to the number of testes, leading into efferent ducts.
5. All these ducts may remain distinct or two of them (when 4 are present) of one side may open into a common átrium or may join to form a single duct, the vas deferens, opening into the male genital atrium.
6. In connection with the terminal part of the vas deferens there is a gland, the prostrate gland.
7. In many earthworms, there are some special setae near the aperture of the vas deferens known as penial setae.
8. There are present four pairs of spermathaecae at the anterior part of the body cavity in which spermatophores are stored in some of the aquatic and terrestrial forms.

Female

1. There are only two small ovaries.
2. The ova become matured in the ovary, or groups of cells may become detached from it and one cell in each group ripens into an ovum.
3. A receptaculum ovarium may also present occassionally.
4. There are two oviducts with a pair of funnel shaped opening into the coelom; generally they open outside through the female genital opening.

5. There is a glandular thickening of the integument, called the clitellum, in one or more segments of the body. The clitellum is saddle-shaped. Its secretion helps to make the cocoon or in some Earthworms, it helps the animal to be bound together during copulation.

Hirudinea

They are hermaphrodite.

5

PHYLUM ARTHROPODA

General Characters and Classification of Arthropoda

Characters

The phylum Arthropoda (*Arthros* – jointed; *podia* – feet) is the largest known phylum in the animal kingdom consisting of animals having jointed appendages, such as Prawn, Cockroach, Scorpion, etc. Members of the phylum can fly, live underground, in water and also on land.

1. The body is bilaterally symmetrical and segmented with three germ layers (ectoderm, endoderm and mesoderm). The segments are variously grouped and fused forming distinct regions e.g., head, thorax and abdomen. The number of segments is limited unlike that of Annelida.

2. Several or all of the segments bear paired jointed appendages which are provided with well-developed musculature and nerve supply. Appendanges are variously modified.

3. The body is covered by a thick chitinous exoskeleton secreted by the endodermis. The exoskeleton is strong and inflexible in most parts and soft at the joints and periodically moulted.

4. The muscles of the body are striated; they are used for rapid action of the body.

5. The body cavity or coelom is reduced in the adult being occupied by the haemocoel which performs a vascular function.

6. The digestive tract is complete; the mouth is ventral and is surrounded by various mouth parts; the fore-gut and hind gut are both long and lined by cuticle continuous with the exoskeleton; the mid-gut is lined by endoderm.

7. The heart is tubular and dorsal to the gut; the blood is colourless; the vascular system is of open type.

8. The respiratory system is well developed; it consists of gills, trachea, lung-books, etc.

9. The excretory system consists of green glands (coxal glands) or Malpighian tubules; nephridia absent except in Peripatus.

10. In the nervous system, the cerebral ganglia are connected by a ring round the gullet to a double chain of ventral nerve-cord which is ganglionated usually in each segment.

11. The sensory system includes many jointed antennae, compound or simpled eyes and sensory hairs, in a few statocysts and auditory organs.

12. The sexes are almost always separate, often associated with sexual dimorphism; genital organs and their ducts are usually paired; fertilization is usually internal; development is direct or indirect with larval forms and metamorphosis; certain forms show parthenogenesis; moultings frequent in young stages; growth by ecdysis.

Classification

The phylum Arthropoda has been divided into the following five classes viz., Crustacea, Insecta, Arachnida, Myriapoda and Onychophora.

Class 1 Crustacea

1. The body is elongated and divisible into head, thorax and abdomen. In most cases the head and thorax are fused forming cephalothorax.
2. Head bears two pairs of antennae and three pairs of jaws in addition to other appendages.
3. The abdomen is usually segmented often with appendages.
4. The appendages are typically biramous and performing manifold functions such as – feeding, cutting, swimming, feeling, etc.
5. The chitinous cuticle is hardened by impregnation with calcium carbonate, thus forming a strong exoskeleton.
6. The coelom is restricted to the gonads as gonocle.
7. Respiration is aquatic and is effected usually by gills.
8. Excretion is by green glands.
9. The sexes are separate or united. Sexual dimorphism is common. Parthenoginesis also occurs. The development is indirect with several larval forms.

Example – Prawn (*Palaemon*), Crabs, Apus, etc.

Class 2 Insecta

1. They are bilaterally symmetrical, triploblastic, metamerically segmented tracheate arthropods.
2. The integument is hardened to form exoskeleton of sclerotized plates.

3. The body is divided into head, thorax and abdomen.
4. The head bears one pair of uniramous antennae, three pairs of mouth appendages and a pair of compound eyes.
5. Thorax has three segments, each segment with a pair of legs ventrally, while dorsally the last two segments have each a pair of wings.
6. Abdomen has no appendages in the adult, and has variable number of segments.
7. Salivary glands are always well developed.
8. Heart is tubular and dorsal and divided into eight chambers, enclosed in the pericardial sac, perforated by paired and segmental ostia. Circulatory system is open, devoid of arteries and veins.
9. Respiration is by branched trachea which convey air to the tissues.
10. The excretory system includes malphigian tubules joined to the hind gut.
11. Sexes separate, development direct or indirect. Usually there is metamorphosis in development.

    Example – Cockroach (Periplaneta), Ants, Termites, Culex, Gryllus, etc.

Class 3 Arachnida

1. The body is divisible into two regions, viz., Cephalothorax of seven fused anterior segments and the abdomen.
2. Cephalothorax bears six pairs of appendages two of which are called chelicerae and pedipalpi and four pairs of walking legs. Antennae are absent.
3. Abdomen is either segmented or unsegmented and usually without appendages.

4. Eyes are simple.
5. Respiration is effected by trachea, book lungs or book gills.
6. Tubular heart is divided into eight compartments.
7. Sexes are separate and the genital opening lies ventrally in the mesosama towards the antrior part. Fertilization is usually internal. Development is mostly direct and there is no metamorphosis.

   Ex. – Scorpion, *Limulus* (king crab), Mites, Ticks, etc.

Class 4 Myriapoda

1. They are tracheate arthropods having elongated body, differentiated into head and trunk but the remaining portion is not distinguishable into regions.
2. The head bears a pair of many jointed antennae, a pair of eyes and two or three pairs of jaws.
3. The trunk consists of a number of similar segments, each segment bearing one or two pairs of jointed legs, each ending in a claw.
4. Respiration is effected by air tubes or trachea.
5. Salivary glands are always present.
6. Excretion is by Malpighian tubules.
7. Sexes are separate. Gonads single with paired gonaducts. Development with metamorphosis.

   Ex.– Scolopendra (Centipede), Julus (Millipede), etc.

Class 5 Onychophora

1. They are terrestrial, with caterpillar-like body, covered with a thin layer of pigmented cuticle.
2. It is divided into rings which do not correspond to the segmental arrangement of the appendages.

3. The head is not well marked off from the trunk.
4. A pair of pre-oral antennae, a pair of jaws, a pair of oral papilae and many pairs of short stumy legs, each with two claws are present.
5. Respiration is effected by means of tracheae, opening into irregularly scattered pits.
6. The heart is elongated and dorsal and with valvular ostia.
7. Cilia are present in the reproductive system.
8. Genital opening is single, median and lying in front of the anus.
9. Almost all are viviparous.

Ex.– Peripatus, Eoparipatus, etc.

PRAWN (PALAEMON)

External morphology

A full grown prawn is nearly 6 to 8 inches long and is pale blue in colour. The elongated body is compressed from side to side, consisting of an anterior, rigid unjointed cephalothorax, and a posterior, movable, jointed abdomen. The body is covered by a cuticle which forms the exoskeleton of the animal. The cephalothorax is formed by the fusion of the head and thorax. It bears 13 pairs of appendages on the ventral surface.

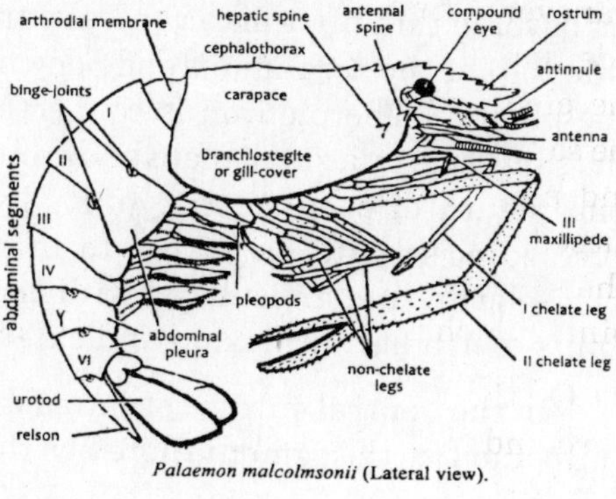

Palaemon malcolmsonii (Lateral view).

The first five pairs of appendages belong to the head and the remaining eight to thorax. The cephalotorax is covered dorsally by a huge shell called the carapace which is a part of the exoskeleton. The carapace is anteriorly prolonged into a saw-like median process called the rostrum. The carapace hangs freely on the sides so as to enclose, on either side, a narrow space called the gill-chamber. The lateral portion of the carapace covering the gills is called the branchiostegite. At the base of the rostrum on each side, there is a distinct orbital notch for accommodating the eye. The two eyes are mounted on movable eye stalks. Below and behind each eye-stalk, there are two small spines of which, the anterior longer one is called the antennal spine and the posterior shorter one is known as hepatic spine.

The abdomen is composed of six distinct segments each bearing a pair of biramous swimming legs called pleopods or swimmerets. At the posterior end of the abdomen there is a median conical piece called telson which does not bear any appendages. The free end of the telson is sharp and pointed, bearing a pair of spines.

Each abdominal segment is covered by a ring-like calcified piece, called sclerite. The successive sclerites are connected with one another by thin membranes called arthrodial membranes which facilitate movement of sclerites. The convex dorsal part of the sclerite is the tergum, while the narrow transverse thin plate-like ventral part is the sternum. Each side flap of the tergum extends downwards on each side as the pleuron. The pleuron is connected to the appendage of the same side by a small plate known as epimeron. The tergum and pleura of each sclerite partially overlap those of the succeeding sclerite, giving rise to an imbricate arrangement. The selerites are articulated with one another by five hinge joints which permit flexion and extension of the abdomen.

On the ventral surface of the cephalothorax between the third and fourth segment there lies the mouth which is a slit-like aperture bounded anteriorly by a plate-like labrum. The anus is a longitudinal slit lying on the ventral surface of the telson near the base.

The male gonopore of each side lies on a papilla on the inner side of the base of the fifth walking leg. The female gonopore of each side lies on the inner surface of the base of the third walking leg. The paired renal apertures are situated internally at the bases of the two second antennae. There are also two minute openings, one on each antennule, by which the statocysts communicate with the exterior.

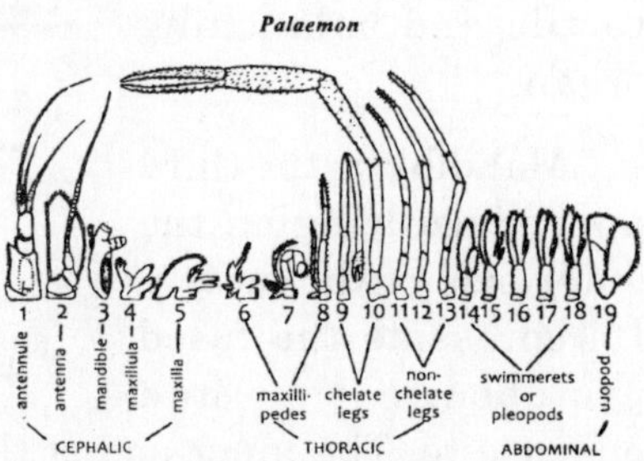

Appendages of *Palaemon* arranged serially.

Appendages of Prawn

The prawn has 19 pairs of appendages, of these 5 pairs are in the head, 8 pairs in the thorax and 6 pairs in the abdomen. An appendage is said to be biramous as it bears two branches or rami. A typical biramous appendage consists of a two jointed stalk of protopodite, the two segments of which are the coxa and the basis; attached to the basis, there are two branches – an outer exopodite and an inner endopodite.

I. **The cephalic (head) appendages**: It consist of the antennules, the antennae, the mandibles, the maxillulae and maxillae. The first antenna (antennule) is situated in front of the mouth, just beneath the eye-stalk. It consists of a three-jointed protopodite, bearing three, many jointed flagella at the distal end and a statocyst at the base. It is sensory and tactile in function.

The second antenna is situated below and behind the first antenna. It consists of a two-jointed protopodite bearing a flat squama and many jointed flagellum. The first segment of the protopodite is coxopodite and the second segment is the basipodite. The coxopodite bears the excretory organ or green gland. The second antenna functions as tactile and balanching organ.

Mandible is the third appendage lying on the outer side of the mouth. It represents the fused protopodite and endopodite. The inner end of the mandible forms a cutting blade provided with teeth, while the middle portion is produced inward into a masticatory lobe (molar process) for grinding food. A three-jointed mandibular palp arises from the outer surface of the mandible. The mandibles serve for cutting and crushing of the food.

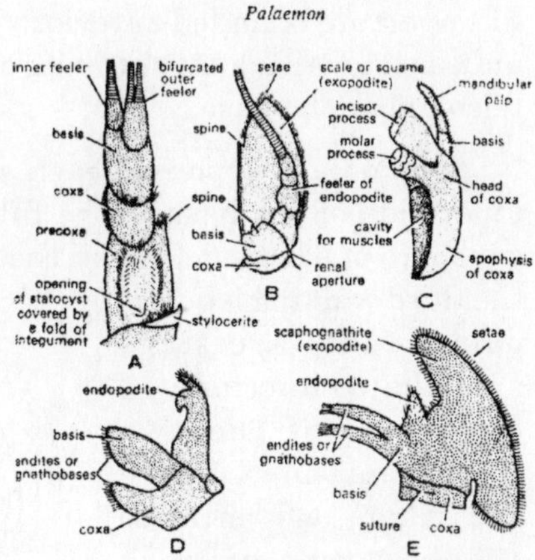

Cephalic appendages of *Palaemon*. A—Antennule. B—Antenna. C—Mandible. D—Maxillula. E—Maxilla.

The first maxilla (maxillula) is the smallest of all the appendages. It lies behind the posterior lip of the mouth. It is leaf-like with two inner lobes acting as gnathobases and an outer lobe which is bifurcated towards the exremity. The maxillula helps in tearing and pushing the food into the mouth.

The second maxilla is also leaf-like. Its protopodite consists of a much reduced coxa, and a large bifurcated basis which projects inward to form a forked gnathobase or jaw. The large fan-shaped exopodite is called the scaphognathite. The endopodite is a short conical process arising from the top of the protopodite. This appendage helps in mastication and respiration.

II. Thoracic appendages

These consist of three pairs of maxillipedes and five pairs of walking legs. The first maxillipede is leaf-like. It has a protopodite consisting of a coxa and a basis. These carry stiff setae on their inner margins. They act as gnathobases which help in pushing the food into the mouth. The outer side of the coxa bears a bilobed leaf-like epipodite. The exopodite is whip-like with a flat lobe at the base. The first maxillipede is respiratory, masticatory in function.

Palaemon

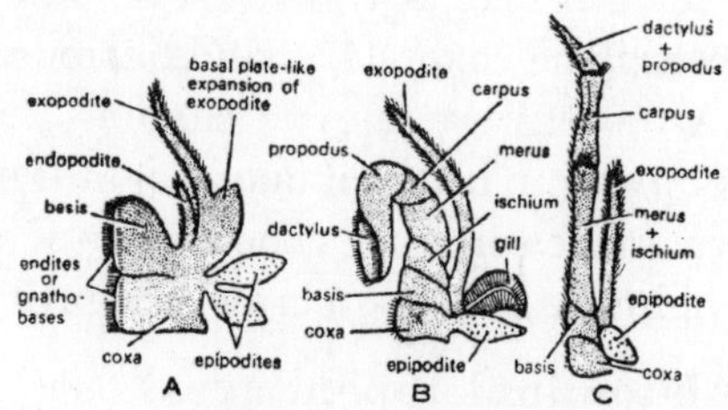

Maxillipedes of *Palaemon*. A—First. B—Second. C—Third.

The second maxillipede has a five-jointed endopodite. The five segments are called ischium, merus, carpus, propodus and dactylus. The protopodite is covered with setae in the innermargin and it carries a small epipodite and a gill on its outer margin. The exopodite is long, slender, unjointed whip-like process. The second maxillipede is respiratory in function.

Arthoropoda

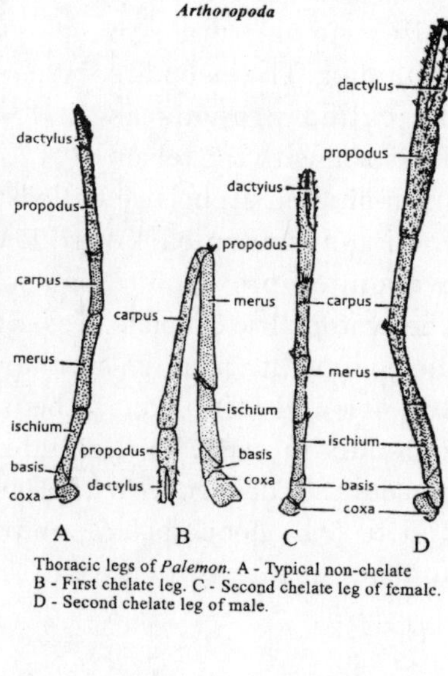

Thoracic legs of *Palemon*. A - Typical non-chelate B - First chelate leg. C - Second chelate leg of female. D - Second chelate leg of male.

The third maxillipede is leg-like in appearance. The protopodite bears a three-jointed leg-like endopodite and a slender whip-like unjointed exopodite. A small epipodite is also present. The third maxillipedes help in locomotion and respiration of the animal.

Walking legs

There are 5 pairs of walking legs ; each leg consists of seven podomeres or segments. The first two legs end in chela and the second is the largest. The two basal segments represent the coxopodite and basipodite, and the remaining five are ischium, merus carpus, propodus and ductylus respectively in order of succession. The chelate legs help the prawn in holding the food before it is taken. The walking legs help the prawn to crawl upon the river bed.

III. Abdominal appendages

There are 6 pairs of abdominal appendages, born on six abdominal segments. The first five pairs are called pleopods or swimmerets. They help the animal in swimming. The last pairs are called uropods as they form with the telson a fan-shaped structure called the tail fin which is used for making a sudden backward leap. Each pleopod consists of a two-jointed protopodite bearing exopodite and endopodite. The protopodite consists of coxa and basis. In the first pleopod, the endopodite is much smaller than the exopodite. The 2nd, 3rd, 4th and 5th pleopods bear rod-like processes known as appendix interna. In the 2nd pleopod of male, an appendix masculina concerned with reproduction, is present. In the 2nd to 5th pleopods, appendix internae are present. The appendices interna of the two sides interlock with one another

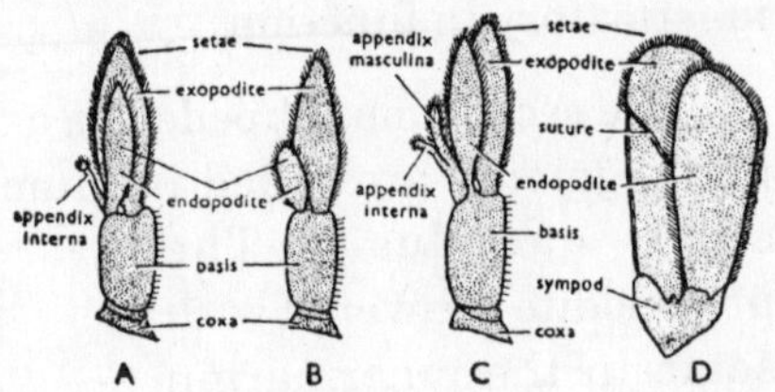

Abdominal appendages of *Palaemon*. A—Typical. B—First. C—2nd of male. D—Uropod.

during the breeding season in female prawn thereby forming a basket-like structure for carrying the eggs. The 6th abdominal appendage, called the uropod has a protopodite of only one segment. The exopodite and endopodite are broad and oval in shape.

Digestive system

The digestive system of Prawn consists of the mouth parts, the alimentary canal and the digestive glands, called the hepatopancreas (liver).

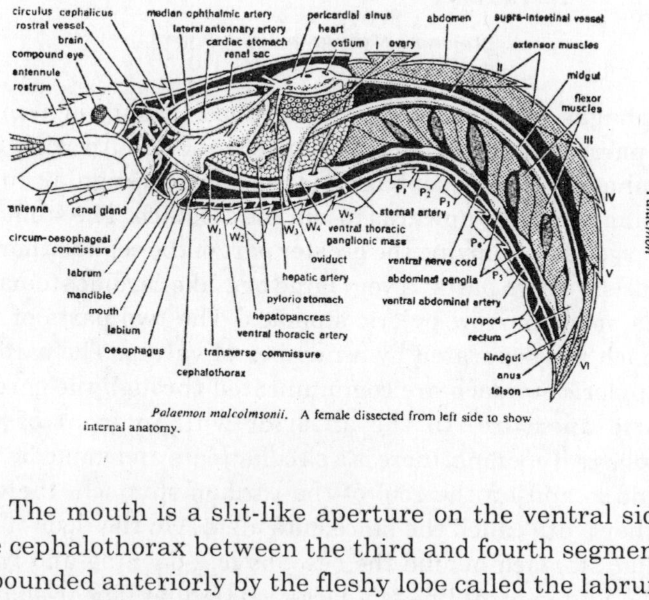

Palaemon malcolmsonii. A female dissected from left side to show internal anatomy.

The mouth is a slit-like aperture on the ventral side of the cephalothorax between the third and fourth segment. It is bounded anteriorly by the fleshy lobe called the labrum or the upper lip and behind by the labium or lower lip, and on each side by the incisor of a mandible.

The alimentary canal extends from mouth to anus and consists of three parts, namely the foregut, the midgut and the hindgut. The foregut and hindgut are lined internally by ectodermal layer, while the midgut is lined by endodermal layer. The foregut consists of mouth, buccal cavity,

Palaemon

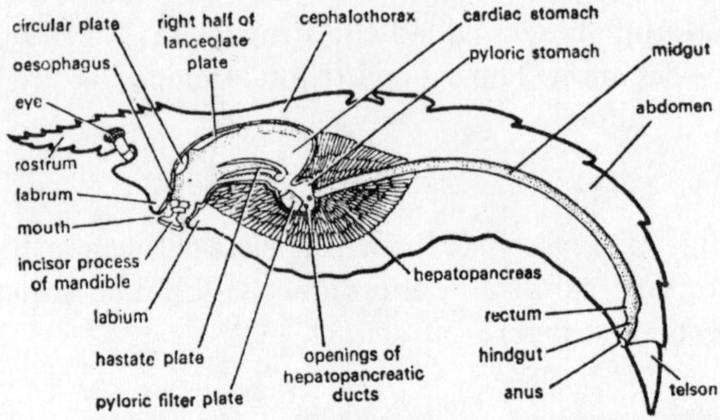

Palaemon. Alimentary canal in lateral view.

oesophagus and the stomach. The buccal cavity is a short and narrow chamber which connects the mouth with the oesophagus. The oesophagus is a wide rectangular tube, running vertically upwards to join the stomach. The stomach is a large sac, occupying the greater part of the cephalothorax. It consists of two parts: a very large bag-like cardiac stomach, and a small narrow pyloric stomach. The two parts of the stomach are separated by a number of valves. The cardiac and pyloric stomach are communicated through the cardio-pyloric aperture. In the anterior wall in front of the oesophageal opening, there is a circular plate and immediately behind i' and on the roof of the cardiac stomach, there is another p ate called the lanceolate plate. On the floor of the cardiac stomach behind the oesophageal opening and lying on the mid-ventral line is a large, triangular, spear-shaped plate called the hastate plate. The broad base of the hastate plate extends posteriorly, forming the anterior margin of the cardio-pyloric opening. On each side of the hastate plate there is a curved plate called the combed-plate. On either side of the hastate plate there is a narrow tunnel called lateral groove. At about the junction of the roof and the floor of the cardiac stomach and outer to the combed-plate, the wall is produced into a longitudinal fold called the guiding ridge which serves to direct the contents of the cardiac stomach towards the cardio-pyloric opening.

The pyloric stomach is a small, narrow and muscular chamber. It receives on each side a duct from the hepatopancreas which is a digestive gland, orange red in colour occupying a large part of the cephalothoracic cavity. The hepatopancreas serves both the functional of liver and pancreas.

Palaemon

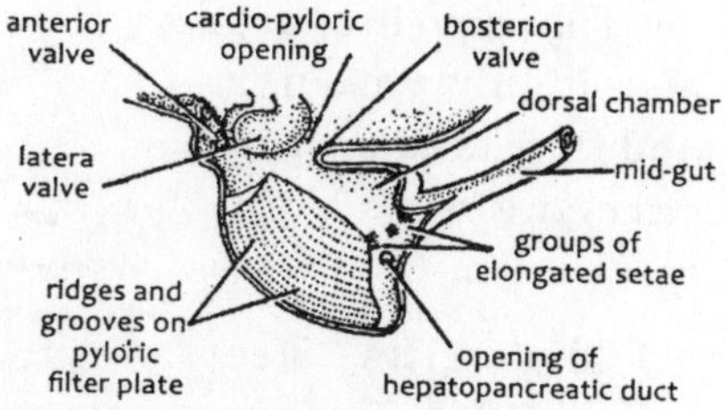

Palaemon. Pyloric stomach after removal of left wall (side view).

The pyloric stomach is followed by the mid gut or intestine which is the longest part of the alimentary canal. It lies in the mid-dorsal line and runs towards the posterior end upto the 6th abdominal segment. The midgut leads into the hind gut which is the shortest part of the alimentary canal. The anterior part of the hind gut is dilated to form an intestinal bulb or rectum which opens to the exterior by the anus on the ventral side of the base of the telson.

Mechanism of feeding and digestion

Prawn feeds on algae, mosses, insects and decaying organic matter. Food is seized by the chelate walking legs and pushed into the buccal cavity by the help of maxillipedes, maxillula and maxillae. The food is crushed by the molar processes of the mandibles. The crushed food is then sucked into the cardiac stomach by alternate contraction and expansion of the oesophageal walls as well as cardiac stomach.

The food is digested by the action of the digestive juice from the hepatopancreas. The digested food is then absorbed in the intestine. The undigested residue is egested through the anus.

Respiratory system

The respiratory organs of Prawn consists of the lining membrane of the carapace, the three pairs of epipodites, and eight pairs of gills or branchial.

The lining membrane of the carapace is thin and vascular and it forms a large respiratory surface through which exchange of oxygen and carbon dioxide takes place.

The epipodites are also highly vascular and they take part in carrying out respiration.

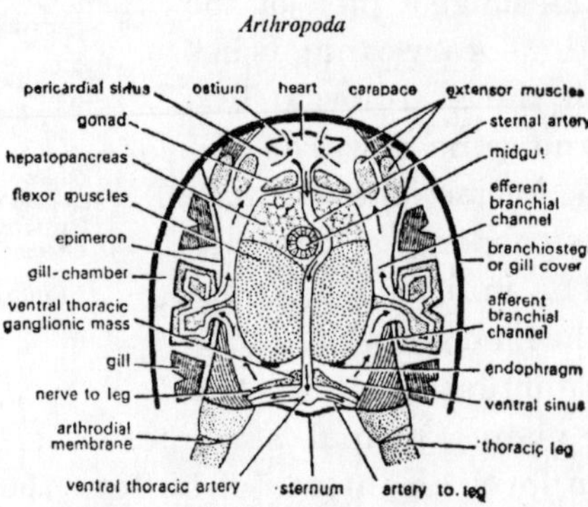

Palaemon. A section through the cephalothorax to show the two gill-chambers.

The gills are arranged on either side of the cephalothorax and covered by branchiostegite. The gills are more or less crescent-shaped and they increase in size antero-posteriorly, so that the first gill is the smallest and last one (7th) is the largest. Each gill consists of a double row of plates containing blood channels which lead back into the pericardium. The gills are closely set like the leaves of a book. On the basis of their place of attachment, the gills may be either podobranch or arthrobranch or pleurobranch. The first gill is attached to the coxopodite of the second maxilliped and is a podobranch. The second gill (arthrobranch) is attached to the arthrodial membrane, connecting the third maxilliped with the body. The remaining five gills (pleurobranchs) are attached to the outer side of the cephalothorax above the articulation of the walking legs.

Mechanism of respiration

The scaphognathite moves constantly so as to produce a current of water. Water enters the gill chamber along the posterior and ventral margins of the branchiostegite and after bathing the gills and epipodites comes out near its antero-dorsal end. The gaseous exchange between the blood and the water takes place by diffusion. Oxygen is taken in by blood and carbon dioxide from blood diffuses out into the water.

Blood vascular system

The blood vascular system of Prawn is of open type. Blood vessels freely communicate with spaces which collectively form a haemocoel. Branches of the arteries or capillaries pour their blood not into veins but into the large spaces between the muscles and the various internal organs.

The blood vascular system consists of the heart enclosed by the pericardium, the arteries or true blood vessels with definite muscular walls, and the irregular sinuses or blood lacunae into which the arteries open directly.

The blood is a clear fluid with colourless leucocytes. In the blood, the plasma contains a pigment called haemocyanin. When oxygenated, the blood appears to be dark blue otherwise it is colourless.

The heart is a muscular triangular organ situated in the mid dorsal line at the posterior part of the cephalothorax. It is enclosed by the pericardial sinus, the thin wall of which is called pericardium. It is held in position by a muscular cord, the cardio-pyloric strand. The muscular wall of the heart is perforated by five pairs of ostia which allow the blood to flow from the pericardial sinus to the heart but not vice versa. Of the five pair is on the ventral surface, the third pair on the posterior border and the remaining two pairs pierce the lateral borders of the heart. The heart is a spongy structure as its cavity is pierced by a large number of muscle fibres running in all directions.

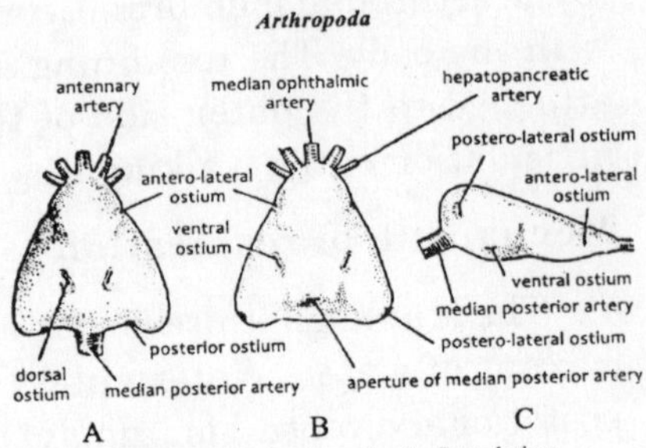

Palaemon. Heart showing ostia. A - Dorsal view. B - Ventral view. C - Lateral view.

The heart receives blood from the pericardial sinus and by repeated contractions, pumps out blood into the arteries which are provided with strong muscular wall.

The main arteries are the following : (1) a median ophthalmic artery, (2) paired antennary arteries, and (3) paired hepatopancreatic arteries.

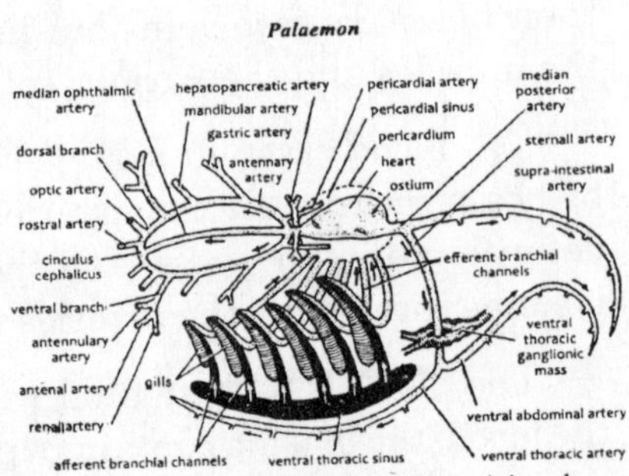

Palaemon. Heart and principal arteries and channels.

The median ophthalmic artery arises from the anterior end of the heart and runs forward upto the roof of the oesophagus, where it ends by anastomosing with the antennary arteries.

The paired antennary arteries arise from the apex of the heart, one on either side of the median ophthalmic. Each antennary artery supplies a pericardial branch to the pericardial sinus, a gastric branch to the stomach, and a mandibular branch to the mandibular muscle. After giving off these branches, each antennary artery continues forward upto the basal segment of the antenna of its own side where it breaks up into a dorsal and a ventral branch. The dorsal branch turns inward and supplies an optic artery to the eye. It then ends by anastomosing with its fellow of the opposite side and with the median ophthalmic. An arterial circle is thus formed anteriorly and this gives off a pair of rostral arteries to the rostrum. The ventral branch of the antennary artery divides into two, one of which, the antennulary artery, supplies the antennule and the other supplies the antenna (second antenna).

A pair of hepatic arteries arise from the ventral surface of the apex of the heart to supply the hepatopancreas.

The posterior artery which is single and median in position arises from the postero-ventral surface of the heart. It divides into two branches, namely a dorsal abdominal or supra intestinal, and the sternal artery. The dorsal abdominal artery is slender, straight and median artery, lying immediately above the midgut. It runs backward upto the hind gut and supplies the dorsal abdominal muscles and the midgut.

The sternal artery is the stoutest of all the blood vessels in prawn. It runs downward and forward, passing either to the left or to the right of the midgut and after piercing the cephalothoracic ganglionic mass comes to lie on the ventral side where it divides into two brachnes, namely the ventral thoracic artery and the ventral abdominal artery. The ventral thoracic artery comes forward upto the mouth and supplies blood to the ventral part of the cephalothorax. The ventral abdominal artery runs backward along the ventral region of the abdomen and supplies blood to the hindgut and the ventral region of the abdomen.

Course of blood circulation

The heart receives blood from the pericardium through

the ostia. From the heart blood enters the arteries and thence into their minute branches which open into spaces called locunae. The lacunae of the body intercommunicate with one another and lead into two large spaces, called the ventral sinuses which are situated in the ventral region of the cephalothorax. Blood from the ventral sinus passes into the gills by a number of afferent branchial channels or veins. The oxygenated blood from the gills returns back into the pericardial sinus by six pairs of efferent branchial channels. From the pericardial sinus, the blood enters the heart through ostia. The heart by systolic contraction, pumps the aerated blood into the arteries. Leakage of blood into the pericardial sinus is prevented as the ostia have valves which permit the blood to flow from the sinus into the heart and not vice versa. The arteries ramify into all parts of the body supplying the tissues with oxygen. The deoxygenated blood is collected in the ventral sinuses and then sent back to the gills for oxygenation.

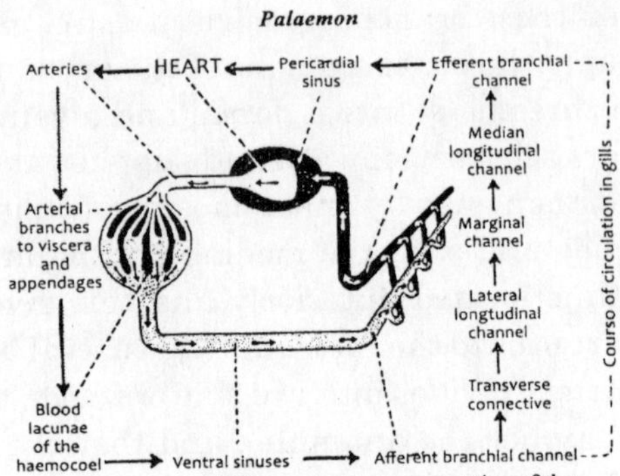

Palaemon. A diagrammatic representation of the course of circulation of blood

Excretory System

The excretory system consists of a pair of green glands or antennary glands lying within the coxa of the antenna of its own side. A green gland has three parts – an end sac, a labryinth or glandular mass, and a bladder. The end sac is bean-shaped, containing a blood lacuna. It lies in between the glandular labryinth and the bladder. Its wall consists of an outer layer of connective tissue containing numerous blood

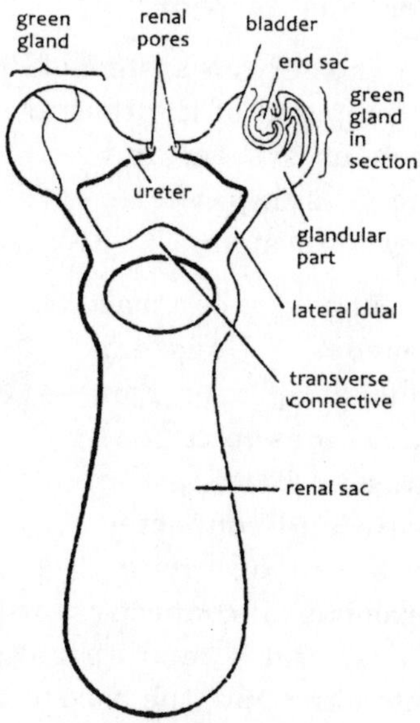

Palaemon. Excretory organs in dorsal view.

spaces, and an inner layer of excretory epithelial cells. The lumen of the end sac opens by a single aperture into the labyrinth. The labyrinth is situated on the outer side of the end sac, and is composed of a mass of excretory tubules opening into the bladder. The bladder is a thin-walled sac lying on the inner side of the end sac. It communicates with the urinary opening by a short ureter which opens to the exterior through the small renal opening, situated on a small elevation on the inner side of the coxa of the antenna of its own side. At the point of junction of the labyrinth and the bladder, the antennary gland is produced into a narrow tubular lateral duct. The two lateral ducts run, one on each side of the oesophagus and meet each other by a transverse connection. The lateral ducts open into the renal sac which lies just above the cardiac stomach and beneath carapace. The renal sac acts as a reservoir, where the excretory products may be temporarily stored. The end sac and the labyrinth form excretory products which are carried to the bladder, and excreted out through the renal pores.

The integument of prawn also acts as an accessory excretory organ as it separates nitrogenous waste products which are stored in the hard outer shell of the body. These are eliminated at the time of moulting.

Nervous System

The nervous system of Prawn is of the annelidan type. It is composed of the central nervous system, the peripheral nervous system and the sympathetic nervous system.

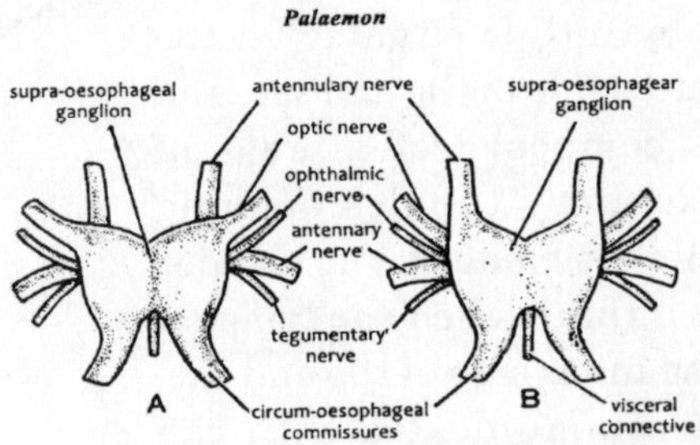

Brain of *Palaemon*. A - Dorsal view. B - Ventral view.

The central nervous system consists of a pair of supra-oesophageal ganglia forming the brain and connected by a pair of circum-oesophageal connectives with a ventral thoracic ganglionic mass, and a median ganglionated ventral nerve-cord extending upto the base of the telson.

The brain is situated dorsally in the anterior region beneath the base of the rostrum, at the junction of the oesophagus with the cardiac stomach. The brain is bilobed and it represents fused mass of several ganglia. The following nerves arise from the brain:–

1. A pair of optic nerves arises from the dorsal surface of the brain, one on each side. Each runs forward and supplies the retinal region of the eye.
2. Behind the optic nerve, on either side arises a slender ophthalmic nerve and supplies the eye muscles.
3. A stout antennulary nerve arises on either side below the root of each optic nerve. It runs forward and becomes divided into branches which supply the different parts of the appendage including the statocyst.
4. A pair of stout antennary nerve arises from the ventral

surface of the brain close to the origin of the antennulary nerves. Each antennary nerve runs obliquely downward and then curving forward enters the base of the second antenna innervating its parts by branches.

5. A small tegumental nerve arises just behind the antennary nerve on either side and supplies the labrum.

The circum-oesophageal commissures or connectives are stout nerve cords that connect the brain with the thoracic ganglionic mass. These cords arise from the hind end of the brain and run one on either side to embrace the oesophagus and meet posteriorly on the ventral side with the ventral nerve chain. The two connectives are joined by a post-oesophageal loop which embranches the oesophagus posteriorly.

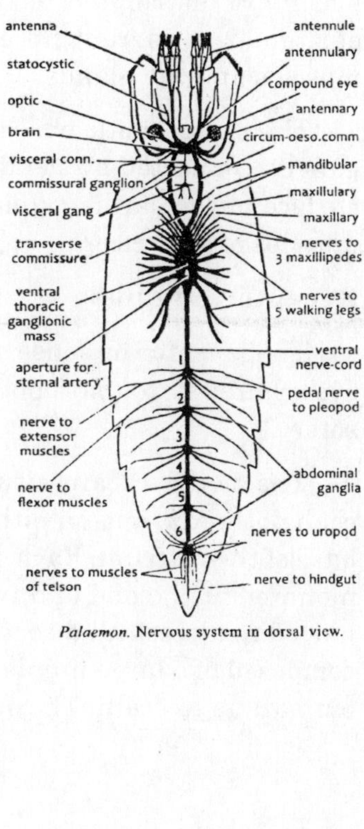

Palaemon. Nervous system in dorsal view.

The cephalothoracic ganglionic mass is an elongated oval structure, lying in the mid-ventral line and is perforated near its posterior end for the passage of the sternal artery. It is composed of a number of ganglia which are fused together. Eleven pairs of nerves arise from the ganglionic mass. These nerves supply all the cephalothoracic appendages excepting the two pairs of antenna.

The ventral nerve cord is composed of two cords fused together. It lies in the mid-ventral line beneath the mass of abdominal muscles. It arises from the posterior end of the cephalothoric ganglionic mass and posteriorly it terminates at the base of the telson. Corresponding to the abdominal segment, the nerve cord gives rise to six ganglionic swellings. Each ganglion supplies the muscles and the other appendages of the abdominal region. The sixth abdominal ganglion is larger than the other. It gives off two pairs of nerves to the uropods, two pairs of nerves to the flexor muscles of its own segment, two pairs of nerves to the telson and an unpaired nerve to the hind gut.

The peripheral nerves arise from the ganglia of the central nervous system. Each nerve is composed of a bundle of nerve fibres. The nerve fibres may be afferent or sensory which are distributed to the receptor organs such as eyes, etc., and the afferent fibres which innervate the voluntary muscles and the glands.

The sympathetic nervous system consists of a few small ganglia connected by slender nerves. These lie on the dorsal surface of the cardiac stomach and supply the involuntary muscles of the viscera.

Receptor Organs

The receptor or sense organs in Prawn consist of paired eyes, paired statocysts or otocysts, tactile organs and olfactory setae.

The two eyes are situated at the anterior end of the cephalothorax, one on either side of the cerapace, near the base of the rostrum. Each eye is hemispherical in shape, and mounted at the end of a two-jointed eye-stalk which can be moved in almost all directions. The eye is compound as it is composed of a large number of simple visual elements called ommatidia or ocelli. The outer or the corneal surface of each eye is convex and is divided into a large number of square facets placed in juxtaposition. Each facet corresponds to a single ommatidium. The ommatidia are arranged along the radius of the hemispherical eye.

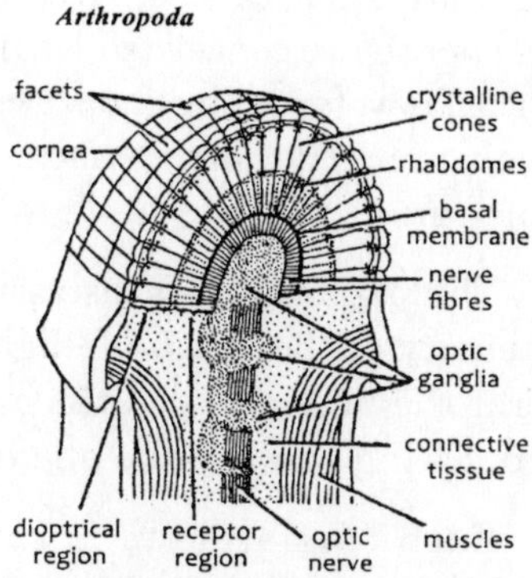

A diagrammatic L.S. of compound eye showing arrangement of ommatidia.

Each ommatidium is a minute, elongated tube consisting of a number of cells placed end to end along a central axis. Just beneath the cuticular corneal facet, there are two flat corneagen cells which secrete a new cornea when the old one is damaged or lost. Beneath the corneagen cells lies a transparent body called crystalline cone or lens which is surrounded by four tall cells, the cone cells or vitrellae. The cone cells secrete and nourish the lens. The crystalline cone is followed by an elongated, spindle-shaped and transversely striated body called rhabdome which is not only secreted but also completely surrounded by seven elongated retinular cells or etinulae. These cells and the rhabdome form the retinal region of the ommatidium, which rests on a basement membrane and serves to form the image. Each ommatidium is optically isolated from the neighbouring ommatidia by dark pigment sheaths composed of chromatophores which are arranged in two groups – a proximal and distal. The proximal group surrounds the rehabdome and retinular cells and is known as the retinal sheath; the distal group surrounds the lens and vitrellae and forms the iris sheath. The retinulae and rhabdome are supplied with fibre-like branches of the optic nerve.

The cornea, corneagen cells, crystalline cone, vitrellae and iris sheath are concerned with focussing of light on the retinal part. They form the dioptrical region of the eye. The light received by the corneal facet is focussed on the retinal part which transmits the same to the brain.

The pigment sheaths can extend and retract as per the intensity of the available light. In dim light and during the night the pigment sheaths are retracted and in bright light and during the day the sheaths are fully extended.

Each ommatidium is capable of producing an image. When the ommatidia are optically isolated from one another by extension of pigment sheaths, composite image is formed being made up of a large number of separate ommatidia. Such an image is formed during the day time and is called a mosaic or apposition image. In dim light or during the night, the image so formed is blurred and indistinct as the pigment sheaths between the ommatidia are retracted and all the ommatidia work together to form a single complete image. Such an image is called a superposition image.

Statocysts

The statocysts or otocysts are paired organs for maintaining balance. Each statocyst is almost spherical cuticular sac lying embedded in the antennules, one to each appendage. The statocyst is attached to the lower surface of the concave roof of the pre coxa and it communicates with the exterior through a small

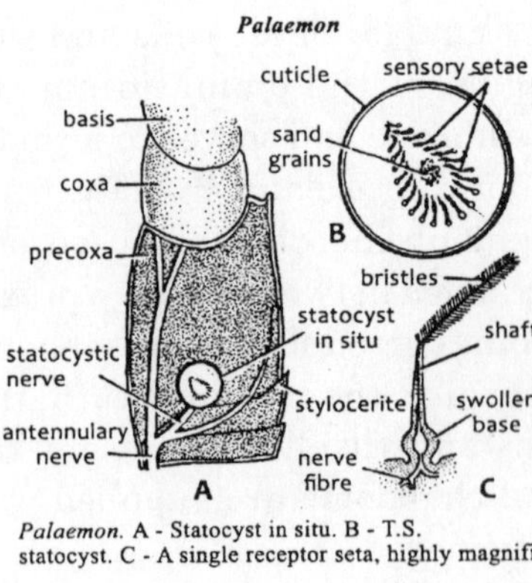

Palaemon. A - Statocyst in situ. B - T.S. statocyst. C - A single receptor seta, highly magnified.

aperture at this place of attachment. Inside the statocyst, there is an oval ring of setae and at the centre a heap of sand grains is present. Each setae is innervated by a fine branch from the statocystic nerve. The sand grains act as statolith. With every change in the position of the animal in water, the sand grains press against the setae, and the stimulus of contact produces an impulse which is conveyed to the brain. The Prawn feels the changed circumstances and quickly re-oreints itself and restores its normal balance.

Tactile Organs

These are concerned with the sense of touch and contact. The many jointed elongated feelers of the antennules and antennae are considered to be tactile in function. The borders of the flattened appendages such as the rami of the swimmerets are fringed with tactile setae. Each tactile seta is a cuticular rod consisting of two segments – a swollen base, and a plumose distal part. The former is attached to the integument by a membrane, while the latter (distal part) is longer and tapering and its free end, and is provided with two rows of barbs.

Olfactory setae

These are concerned with the perception of smell. Between the two long feelers of each of the first antenna there is a small middle feeler which bears small seate. Each seta has two segments of which the proximal is attached to the integument by a flexible membrane and the distal is blade-like. The setae are supplied with antennulary nerve and are considered to have an olfactory function.

Reproductive System

In Prawn sexes are separate. The males are larger in size than the females. The second walking legs in males are larger than that in females and also covered with greater number of spines. The second pleopod in male bears an appendix masculina but the female does not have it. In the male, the paired gonopores are situated at the bases of the fifth walking legs. In female Prawn, the gonopores are situated at the bases of the third walking legs.

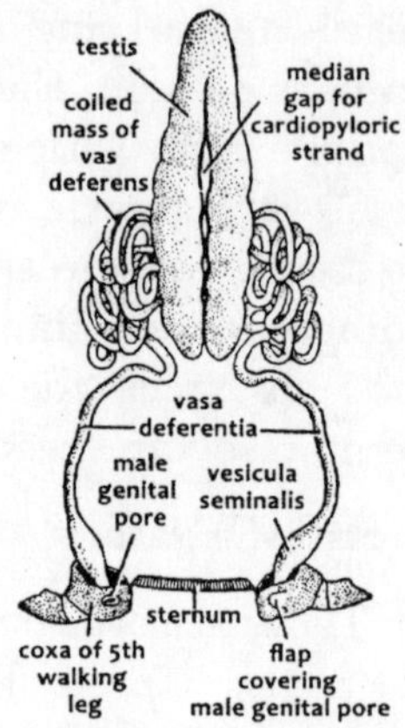

Palaemon. Male reproductive system (dorsal view).

The male reproductive organ consists of a pair of testes, a pair of vasa deferentia and a pair of vesicula seminalis.

The testes are soft and elongated whitish bodies which fuse at the anterior end to form a common lobe. The testes lie above the hepatopancreas and below the pericardial sinus and extend from the renal sac to the first abdominal segment. Each testis is composed of a large number of lobules held together by connective tissue. Spermatozoa are budded off from the inner wall of the hollow lobules.

From the outer side of each testis arises a vas deferens. At its commencement, the vas deferens forms a highly convoluted mass which lies on the dorsal side of the hepatopancreas. Beyond this mass it straightens up and extends backwards and downwards to the base of the fifth walking leg of the same side where it dilates to form a club-shaped vesicula seminalis which opens to the exterior by a male genital aperture. The male genital aperture (gonopore) is paired; each gonopore is situated at the base of the fifth walking leg of its own side. The spermatozoa are carried by the vasal deferentia into the seminal vesicles where they are stored in the form of a compact pockets called spermatophores.

The female reproductive organs consist of a pair of ovaries, and a pair of oviducts. The ovaries are large, brown sickle-shaped bodies which are united at both the ends, leaving a gap in the middle for the passage of the cardio-pyloric strand. The ovaries lie on the dorsal surface of the hepatopancreas and just beneath the pericardial sinus of the heart. Each ovary is enclosed within a membranous capsule and each consists of a very large number of radically arranged rows of ova. The ova are large nucleated cells containing yolk. The shape and size of the ovary may vary according to age of the individual.

From the middle of the outer border of each ovary there runs a short wide tube, the oviduct. This oviduct opens on the inner side of the coxopodite of the third walking leg of its own side.

The testes are soft and elongated whitish bodies which fuse at the anterior end to form a common lobe. The testes lie above the hepatopancreas and below the pericardial sinus and extend from the renal sac to the first abdominal segment. Each testis is composed of a large number of lobules held together by connective tissue. Spermatozoa are budded off from the inner wall of the hollow lobules.

From the outer side of each testis arises a vas deferens. At its commencement, the vas deferens forms a highly convoluted mass which lies on the dorsal side of the hepatopancreas. Beyond this mass it straightens up and extends backwards and downwards to the base of the fifth walking leg of the same side where it dilates to form a club-shaped vesicula seminalis which opens to the exterior by a male genital aperture. The male genital aperture (gonopore) is paired; each gonopore is situated at the base of the fifth walking leg of its own side. The spermatozoa are carried by the vasa deferentia into the seminal vesicles where they are stored in the form of a compact pockets called spermatophores.

The female reproductive organs consist of a pair of ovaries, and a pair of oviducts. The ovaries are large, brown sickle-shaped bodies which are united at both the ends, leaving a gap in the middle for the passage of the cardio-pyloric strand. The ovaries lie on the dorsal surface of the hepatopancreas and just beneath the pericardial sinus of the heart. Each ovary is enclosed within a membranous capsule and each consists of a very large number of radically arranged rows of ova. The ova are large nucleated cells containing yolk. The shape and size of the ovary may vary according to age of the individual.

From the middle of the outer border of each ovary there runs a short wide tube, the oviduct. This oviduct opens on the inner side of the coxopodite of the third walking leg of its own side.

6

NERVOUS SYSTEM OF MOLLUSCA

In all but the most lowly organized animals there is a nervous system by means of which the communication is affected between the various parts of the body, and by means of which also a communication is established with the external world, enabling the creatures to work in harmony.

Although arranged in the same general plan, the nervous system in Mollusca presents a good deal of variations in different classes and in different groups of the same class.

Basic parts of the Molluscan nervous system

1. A circum-oesophageal ring.
2. A pair of cerebral ganglia.
3. A pair of pleural ganglia, lying at the sides of the pharynx.
4. A pair of pedal ganglia.
5. A pair of visceral ganglia lying posteriorly into the body cavity.
6. Nerve cords connecting these ganglia and originating from them to supply different organs.

Giving the name "connectives" to such nerves as unite the ganglia of one side of the body that is, dissimilar ganglia, and that of commissures to the nerves that unite the similar ganglia of the two sides of the body, we have the following system:

Commissures are found – (i) between the two cerebral ganglia (over the foregut); (ii) between the two pedal ganglia (under the foregut); (iii) between the two visceral ganglia (under the hind gut). The connectives on each sides are (i) the cerebro-pedal, (ii) the cerebro-pleural, (iii) the pleuro-pedal, and (iv) the pleuro-visceral connectives.

Embryology of the nervous system

The various pairs of nerve centres in Mollusca arise separately, and usually as thickening of the ectodermal at the points where they are formed. In certain cases, however, the nerve centres are developed by the more primitive process of invagination, as for instance, the cerebral ganglion of Dentalium (Lankester E. Ray).

Nervous system in the different classes of Mollusca

Class I Solenogastres

1. The nervous system consists of four longitudinal nerve-cords, two pedal and two pleural. These are connected together by an oesophageal ring, thickened dorsally into a single or double cerebral ganglion; and in front of this is a second, more slender buccal (stomatogastric) nerve ring with small ganglion.

2. The pedal cords may present in front a pair of ganglionic thickenings connected by a commissure, and further back there may be a series of enlargements united by commissures.

3. The pleural cords which are more or less enlarged in front into ganglionic swellings, are connected behind, above the rectum, by a commissure which bears a median enlargement. Sometimes a union takes place posteriorly between the cords of the two pairs.

Variation

The cerebral ganglia in Neomenia are not very prominent and are only formed by a thickening of the oesophageal ring. It may be single or double.

In Chaetoderma, due to the disposition of various parts of the organism the pleural ganglia are placed above the cerebral, and the commissures between the pedal cords are absent.

Class II Placeophora

The nervous system is very primitive. Ganglionic cells are not localised in definite ganglia.

1. There is an oesophageal nerve ring consisting of a thicker dorsal cerebral portion, not differentiated into ganglia, and a thinner ventral commissure. The cerebral part sends off nerves to the labial palps, the lips and to the buccal nerve ring with its ganglionic swellings. The ventral commissure gives off nerves to the sub-radular ganglia.
2. Four medullary nerve-cords run backwards : (a) one pair lateral or pleuro-visceral cords, and the other (b) pedal cord. The visceral and pedal are connected anteriorly and on opposite side by transverse commissures. The pleuro-visceral cords form the visceral loop. The pleuro-visceral cord sends off nerves chiefly to the mantle and ctenidia, while the pedal gives off nerves to the foot.
3. The buccal ganglia are on the buccal ring, connected with the oesophageal ring. This oesophageal ring is not differentiated into distinct ganglia.
4. In Chiton, the buccal and sub-radular ganglia are present but other major ganglia like cerebral, pedal, pleural ganglia are absent.

Remark on the nervous system of Solenogastres and Placophora

The nervous system of these two classes are slightly

"Amphinuran" (hence grouping of these two classes into the one class Amphinura- by some). The system is considered to be primitive type of nervous system in that there is no visceral ganglia and other ganglia are also primitive.

Class III Gastropoda

The nervous system of this class is very important to the comparative anatomists on account of the symmetry and asymmetry of the nervous system in its groups.

General Plan

1. There are two cerebral ganglia near or above the oesophagus, which are connected by cerebral commissures.
2. There are two pedal ganglia below the oesophagus, connected with each other by a pedal commissure, and with the cerebral ganglia by two cerebro-pedal connectives.
3. Two pleural ganglia are connected by cerebro-pleural connectives and pleural commissures.
4. A simple or a complex visceral ganglion lying below the intestine, united to the pleural ganglia by two pleuro-visceral connectives.
5. A parietal ganglion almost always occurs in the course of each pleuro-visceral connective. The parietal ganglion divides the connectives into an anterior pleuor-parietal and a viscero-parietal connective.

Modifications in the groups of Gastropoda

1. In Prosobranchia and Acetaon, the pleuro-visceral connectives cross one another (asymmetrical). The connective springing from the right pleural ganglion runs over the intestine to the left before joining visceral ganglion; while the connective from the left pleural

ganglion runs under the intestine to the right side of the body.

In consequence of this crossing, the right parietal ganglion becomes the supra-intestinal ganglion and lies on the left side of the parietal ganglion and the left side becomes the infra intestinal ganglion which lies on the right side. This type of nervous system is known as "streptoneurous" nervous system.

(a) Sub-order Aspidobranchia (Diotocardia)

1. Nervous system is little concentrated.
2. The pedal ganglia are produced into long ends with the anterior ends of which the pleural ganglia are fused.
3. The cerebral ganglia are wide apart.
4. There is only one indistinct visceral ganglion which is joined with the pleural ganglia by two pleuro-visceral connectives.
5. In Haliotis and Fissurella, the two elongated pedal cords are connected by transverse commissures. The ganglia innervating the osphradia are also found.
6. In Neritidae and Halcinidae the supra-intestinal pleuro-visceral connective and its corresponding ganglia are wanting.
7. In Patella, the pleural and pedal ganglia are joined by a distinct pleuro-pedal connective.

(b) Sub order Pectinibranchia (Monocardia)

1. Nervous system is concentrated.
2. Parietal ganglia are always present.
3. Labial commissures are wanting (except Paludinidae and Ampullaridae).

4. Pedal cord and transverse commissure are wanting (except Architaenioglossa).

5. The number of visceral ganglia varies from 1 to 3-5.

6. "Zygoneura" and "Dialyneura" are remarkable.

7. Development of a new untwisted accessory pleuro-intestinal connective on the right side is said to be a "Zygoneury". When zygoneury takes place on the left side and the pallial anastomosis exists on each side, the nervous system is called "Dialyneuron", example, Cassidaria. Figure (Cassidaria) illustrates "dialyneury" on the left side and "Zygoneury" on the right side. The pallial nerve from left pleural ganglion anastomoses with the pallial nerve of the supra-intestinal ganglion (Dialyneury) ; whereas the pallial nerves from the right pleural run to the infra-intestinal ganglion (Zygoneury).

Origin of the crossing of the pleuro-visceral connective (Chiastoneury)

The asymmetry, in the nervous system of many Gostropods has been explained by the supposition of the fact that the "pallial complex" must have changed its position. It shifted gradually forward along the right mantle furrow, such that each ctenidium would drag along with the pleural ganglion and will develop the "Chiastoneury". (Fig. ABCDE)

2. Order Opisthobranchia

The nervous system of this order in which the typical Gastropodan ganglia are developed, is further characterised by (i) the absence of "Chiastoneury", and (ii) by a marked tendency to the concentration of the ganglia around the posterior end of the pharynx.

(a) In the sub order Tectibranchia, for example, Aplysia, the connection between the visceral ganglia and the supra-intestinal (primitively right) is very short. The infra-intestinal (primitively left) parietal ganglion has disappeared.

In Bulla, the pleural ganglion is shifted ; commissures are comparatively long ; parapodia are innervated from the pedal ganglia.

(b) In the sub order Nudibranchia also the system is characterised by very great construction of the typical molluscan ganglia (e.g. Tethys).

3. Order Pulmonata

The nervous system is "Euthyneurous" owing to the shortening of connectives and concentration of ganglia into a circum-oesophageal ganglionic complex. In Limnaea a pair of cerebral ganglia overlie the oesophagus, and below it is a mass of ganglia in which are to be made out a pair of pedal ganglia and two pairs of ganglia representing the pleural and parietal ganglia. There is also a visceral ganglion. Pedal commissures are often double.

Special remark on the nervous system of Gastropoda

(a) Areas of innervation of the various ganglia

1. The cerebral ganglia innervate the eyes, the auditory organs, the tentacles, the snout or proboscis and buccal mass, and the body walls lying at the base of the snout.
2. The pedal ganglia supply nerves to the musculature of the foot, and also to the columellar muscle.
3. The pleural ganglia send nerves chiefly to the mantle, the columellar muscle and the body wall lying behind the head.
4. The parietal ganglia innervate the ctenidia and osphradium.
5. The visceral ganglia innervate the muscles of the

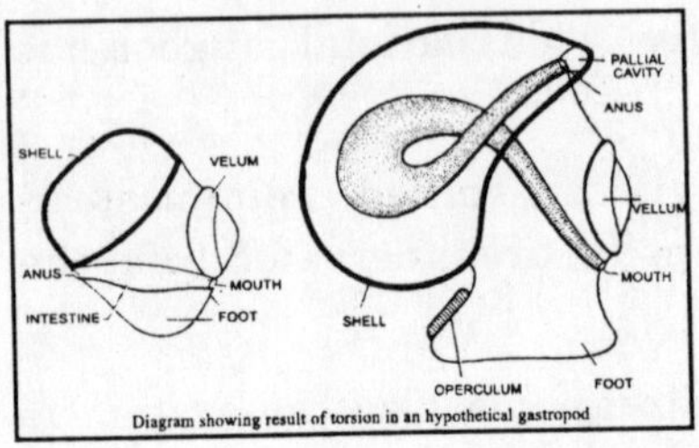

Diagram showing result of torsion in an hypothetical gastropod

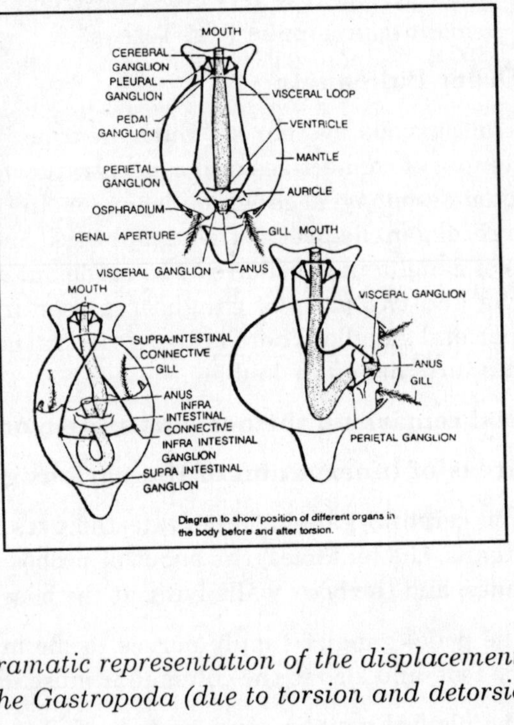

Diagram to show position of different organs in the body before and after torsion.

Fig. *Diagramatic representation of the displacement of nerves in the Gastropoda (due to torsion and detorsion)*

pharynx, the salivary glands, the oesophagus, the anterior aorta, etc.

(b) A step to show that the nervous system of this Gastropoda is derived from Amphinuran nervous system, or the nervous system of Gastropoda is higher than that of the Amphinuran.

1. The cerebral ganglia of the Gastropoda correspond with the oesophageal ring of Chiton, with the exception of the central portion of its lower half, and further with the cerebral ganglia of the Solenogastres.
2. The pedal ganglia of the Gastropoda answer to the pedal cords in the Amphinurans, concentrated each into a single ganglion. The arrangement in the Diotocardia (more primitive prosobranchia), is very interesting. In Diotocerdia the pedal ganglia are continued posteriorly as two true pedal cords, which like those of the Amphinuran, are connected by transverse commissures.

Though it is difficult to compare the pleural, parietal and visceral ganglia of the Gastropoda with Amphinuran, the satisfactory proof seems to be that this whole complex of ganglia, together with its connectives, correspond with the pleurovicceral cords of Chiton.

3. If this last assumption is correct, the pleural ganglion must be supposed to have arisen by the concentration into one ganglion of the part of the pleurovisceral cord of Chiton which contains the pallial ganglionic cells ; this concentration having taken place at the anterior end of the cord, where it leaves the oesophageal ring.

IV. Scaphopoda (Example, Dentalium, Pulsellum, etc.)

1. The nervous system is symmetrical.
2. The visceral commissures are not crossed.
3. The cerebral ganglia lie near one another.
4. The two pedal ganglia lie close to one another on the

anterior side of the foot and joined to cerebral ganglia by two long cerebro-pedal connectives.

5. The two pleural ganglia lie close to cerebral ganglia, so that the cerebro pleural connectives are very short.

6. There are also two visceral loops.

Remark

The nervous system of Scaphopoda is aloof from the types described before, on the other hand, they show a close relationship to the next class just following below.

V Lamellibranchiata (Bivalvia)

1. The nervous system consists of three pairs of ganglia viz., cerebro-pleural, pedal and viscero-parietal.
2. The connectives joining these three pairs are very long.
3. The pedal ganglia lie close together. They supply nerves to the foot.
4. The cerebro-pleurals have arisen by fusion of cerebral and pleural, and are supra-oesophageal in position. They are in contact with the anterior abductor muscle. They supply nerves to the labial palps and the anterior part of the mantle.
5. The viscoro-parietal ganglia supply nerves to the ctenidia, the posterior portion of the abductor muscle and viscera.

Example – Union, Pecton, Anodonta, Nucula, etc.

In Nucula, the nervous system very primitive. Instead of the single cerebro-pleural ganglion of the Anodonta, there are on each side cerebral and pleural ganglia, each united with the pedal by a connective.

VI Cephalopoda

The nervous system of cephalopoda is marked by the

symmetrical nature of the great concentration of the typical molluscan ganglia, including those of visceral connectives.

Sub Class A. *Belemnoidea (Dibranchiata)*

1. The cerebral, pedal, pleuro-visceral are all relatively of large size, closely aggregated together round the oesophagus, supported by cranial cartilage.
2. Cerebral ganglia are fused together in a round mass.
3. Optic ganglia are placed together in the cerebral ganglia.
4. Pedal ganglia are united together in a single mass. Orally this is prolonged and expanded into a broad mass from which ten brachial nerves are given off to the arms. Each brachial nerve expands at the base of the arm into a brachial ganglion. The pedal ganglia also give off nerves to the funnel, and also a pair to the statocyst.
5. The pleuro-visceral ganglia are also united together into one in immediate contact with the pedal ganglida.
6. Besides the optic nerves, the cerebral ganglion gives off two small ganglia.
7. Pleuro-visceral cord gives off numerous pallial nerves.
8. Besides the branchial nerves, there is a branchial ganglion.
9. In Sepia and Loligo, the pallial nerves run to the inner surface of the mantle cavity, where each expands into a large, flat stellate ganglion. From the outer edge of this ganglion, arise a number of nerves supplying the various parts of the mantle.
10. A loop of sympathetic nerves also arises from the inferior buccal ganglia, and runs along the oesophagus to the stomach where it ends in the gastric ganglion.
11. In Octopus, ganglia are much less sharply marked off.

Sub Class B. *Nautiloidea (Tetrabranchiata)*

1. In Nautilus there is no distinct ganglia in the central part of the nervous system.
2. In most cases, they are represented by thick band of nerve collars.
3. The anterior part of the collar represents the cerebral ganglia, the oral portion of the posterior part, the pedal, the aboral portion, the pleuro-visceral.
4. From the cerebral ganglia pass nerves to the buccal mass, olfactory organs and statocyst, and a pair of very thick optic nerves to the eyes.
5. Pedal ganglion gives off numerous nerves to the tentacles and the funnel.
6. From the pleuro-visceral arise pallial and visceral nerves.

7

PHYLUM ECHINODERMATA

Characters and Classification

Characters

1. The echinoderms are exclusively marine animals. They are found at all depths and in all climates. The bilateral symmetry of the larva is replaced in the adult by more or less radial symmetry.
2. They vary in shape, some being rounded and others pentamerous. The body does not show any segmentation, and there is no head.
3. They are triploblastic and coelomate.
4. They are uncephalized and have an oral-aboral axis.
5. Body surface is covered with an exoskeleton of calcareous plates or ossicles, which support a system of movable or immovable calcareous spines.
6. The skin of the animal consists of an outer epidermis and an inner dermis. The musculature lies adjacent to the dermis.
7. The body cavity or coelom is large and it is separated into a large ciliated perivisceral body cavity filled with fluid

containing amoebocytes, and the water vascular system (hydrocoele) comprised of the ring vessel and radial canals communicating with many tube feet, for locomotion, respiration and excretion. During development, the coelom gives rise to haemal and ambulacral systems.

8. They usually have a complete digestive system with mouth but with or without anus.

9. There is no definite excretory system.

10. Respiration takes place through gills or papilae protruding from the coelom by tube feet and in some by cloacal respiratory trees.

11. They have a simple circulatory or haemal system, which runs through a system of coelomic sinuses. The circulatory system is described as open or lacunar type.

12. The nervous system is of diffuse type, uncentralized, typically composed of three nerve rings, arranged around the digestive tube, from which radiating peripheral nerves arise.

13. The sexes are separate and not externally discernible. Reproductive organs are simple with simple gonads. Some are viviparous and some reproduce asexually by self division. The sex cells are generally discharged into the water where fertilization takes place. The larvae are pelagic and they swim by means of ciliated bands.

14. Most echinoderms have the power of casting off and regenerating portions of their body.

Classification

The Phylum Echinodermata is divided into two sub-phyla:

Sub-phylum I *Eleutherozoa*

1. The body is free, without stalk, and is usually placed with the oral surface downwards or in the direction of forward locomotion.

2. A system of ambulacral areas or grooves radiate out from the mouth and contain a double series of tubular appendages, the tube feet ending in suckers.

3. The mouth is directed downwards or forwards and the anus is usually on the aboral surface.

The sub-phylum Eleutherozoa is divisible into five classes:

Class I *Asteroidea (Star Fish)*

1. Body is star-shaped or pentagonal, more or less flattened at right angles to the main axis of the body. The body consists of central disc, from which radiate five tapering arms. The arms are hollow, each arm containing an open ambulacral groove on the ventral surface, with 2-4 rows of tube feet.

2. The oral or aboral surfaces are distinct.

3. The skin has regularly disposed calcareous, often opinous plates.

4. The mouth is directed downwards or forwards. The stomach is large. The anus is on the aboral surface of the central disc. The madreporites is also situated on the aboral surface.

5. The sexes are separate. The larva is either a Bipinnaeria or Branchiolaria.

Ex.– Asterias (Star fish), Solaster (Sun star), Rentaceros, etc.

Class 2. *Ophiuroidea (Brittle Star)*

1. The body is star-shaped with a central disc and with five long, slender flexible arms, which are occasionally branched.

2. The arms are sharply set off from the central disc and they do not contain prolongations of the coelom and

ambulacral grooves. The centre of each arm contains a series of disc-like ossicles which articulate with one another by ball and socket joint. Movement is effected by muscular jerks of arms.

3. The oral and aboral surfaces are distinct.
4. The mouth and madreporite are situated on the oral surface. The stomach is sac-like ; anus and intestine are absent, egestion occurring through the mouth.
5. The tube feet are reduced to sensory papillae.
6. The gonads are pentamerous. The larva is Pluteus (or Ophiopluteus).

Ex.– Ophioglypha (Brittle star), Neoplax, Astrophiura, Ophioderma, etc.

Class 3. *Echinoidea (Sea-urchin, Cake and Heart Urchin)*

1. The body is ovoidal, globular or discoidal, enclosed in a shell of closely fitting, firmly united calcareous plates. The endoskeleton shell also bears relatively long spines which are employed in locomotion.
2. Five double rows of tube feet have terminal suckers disposed penta-radially from the oral surface to the aboral pole of the animal.
3. Alimentary canal is coiled. The mouth is at the lower pole and the anus is either at the aboral pole or in the posterior inter-radius.
4. An intricate set of jaws, known as Aristotle's lantern is present. These jaws are strong enough to chew up the cases of sessile ectoprocts and calcareous algae, and to bore in rocks.
5. The madreporite is situated very close to the anus.
6. There are no ambulacral grooves but the surface is divided

into alternate ambulacral and interambulacral zones running from pole to pole.

7. There are five gonads, supported by mesenteries, and are inter-radial in position.
8. They are dioecious. The larva is Pluteus.

Ex. – Echinus (Sea urchin), Salmacis, Echinocardium, etc.

Class 4. Holothuroidea

1. The body is cylindrical, cucumber-shaped, elongated in the direction of the main axis, with more or less tendency to bilateral symmetry.
2. The skin is leathery with spicules and without spines and without pedicellariae. The dermis encloses endoskeletal ossicles. Musculature is well developed comprising circular and longitudinal muscles.
3. The mouth lies at one end and the anus at the other end. The mouth is surrounded by a circlet of tentacles (modified tube feet, often called buccal podia). Digestive tube is looped, and cloaca is with respiratory tree.
4. Tube feet are distributed over the body, with more on the bottom or ventral surface.
5. Ambulacral grooves are closed. Except a few, there is no external opening of madreporite canal.
6. Respiration takes place through respiratory trees (outgrowths of the cloaca).
7. The gonad is formed of a single or paired tuft of tubules. The larva is an Auricularia.

Ex. – Holothuria, Cucumaria, Synapta, etc.

Sub-phylum *II Pelmatozoa*

1. They are usually fixed echinoderms with a jointed stalk made up of a row or rows of ossicles.

2. They possess slender arms having narrow ciliated ambulacral grooves.
3. The viscera is closed in a calcified and plated theca.
4. The mouth is on the free surface, near the centre, the anus is on the oral surface.
5. From the mouth extends a system of radially arranged, narrow ambulacral grooves, functioning as food-grooves.
6. The circumoesophageal water ring is directly connected with the exterior.
7. The tube-feet are replaced by a small tubular, highly ciliated tentacles which are devoid of suckers.

The sub-phylum Pelmatozoa is divided into five classes:

Class 1. Crinoidea

1. They are usually stalked forms, with five jointed, often branched arms growing out from a central cup or theca, and bearing pinnules.
2. The oral surface is directed upward, and the ambulacral grooves are open.
3. The central parts of the ambulacral, nervous and reproductive systems and the coelom are lodged in theca and send out extensions through the arms.
4. Tube feet in the ambulacral groove have no terminal suckers and internally are without ampullae.
5. The anus lies on the oral surface.
6. There is a water vascular system with connection to the exterior.

Ex. – Antedon (Feather star), Sea lilies, etc.

Class 2. Cystoidea

The class includes extinct forms of lower Silurian to Carboniferous.

The body is ovate or irregular, sessile or shortly stalked and covered with polygonal plates.

Ex. – Carycrinites.

Class 3. Blastoidea

The class includes extinct forms of upper Silurian to Carboniferous.

The body is ovate with five ambulacral areas, with each groove of which jointed pinnules are associated.

Ex. – Pentremites.

Class 4. Edriasteroidea

The class includes extinct forms of Combrian to Carboniferous.

The theca is sac-like with an indefinite number of irregular plate, devoid of any appendages. The mouth is at the centre of the upper surface with five food-grooves bordered by covering plates.

Ex. – Stromatocysties.

STAR FISH

The starfish belongs to the Phylum Echinodermata, sub-phylum Eleutherozoa. Class Asteroidea and order Forcipulata. It is a marine animal, found in the Indo-Pacific ocean and around the West India.

External Features

The body is thick and star-shaped consisting of a central disc from which arise five triangular arms or rays. On the lower or oral surface of the central disc is situated the mouth, and in the centre of the convex aboral surface is situated a minute opening, the anus. The mouth is surrounded by a membranous peristome and by oral spines. On the aboral surface there is a sieve-like plate, the madreporite which is perforated by numerous pores for entry of sea water. Radiating from the actinostome (mouth) into the five arms are five narrow ambulacral grooves. On either border of the ambulacral groove there are two or three rows of movable calcareous spines, the ambulacral spines. The ambulacral groove also contains four rows of tube feet or podia. The distal end of each tube feet is modified into a terminal disc or sucker, which works in the manner of a suction cup. The starfish crawls on its under-surface by applying tips of the tube feet to rocks under the sea. At the distal end of each ambulacral groove there occurs a single terminal tentacle, at the base of which there is a small red spot, the eye spot.

The aboral surface of the starfish also bears a number of white spines, arranged more or less in rows on the arms and supported by ossicles buried in the integument. Between the ossicles there are numerous minute dermal pores, through which project dermal branchiae (gills). Occurring in the spaces between the spines or in clumps around the bases of the spines are minute forcep-like pedicellariae. Each pedicellaria is made up of a pair of blades like those of scissors with a stalk attached to the skin. The blades can be opened and closed by the contraction of minute abductor and adductor muscles, respectively. The pedicellariae may be of two types, the crossed and the uncrossed, according to the disposition of the blades. The crossed pedicellaria occur at the bases of the white spines. The uncrossed pedicellariae are found among the dermobranchiae or dermal gills. The pedicellaria keep the surface of the skin clean and they are also used for seizing the foreign objects.

Body Wall

The body wall consists of epidermis, dermis, muscular

layer, and parietal peritoneum. The epidermis is composed of ciliated columnar cells which extend over all external appendages such as spines, pedicellariae, tube feet, gills, etc. The epidermis also contains neurosensory cells, pigment granules and the gland cells. Lying below the epidermis is a nerve net which is made up of nerve cells and nerve fibres. Beneath the nerve net lies a delicate basement membrane. The dermis lies below the basement membrane. Dermis is the thickest layer of the body wall. It is composed of fibrous connective tissue. The connective tissue is involved in the formation of the calcareous ossicles that together compose the endoskeleton of the animal. Below the dermis is a layer of smooth muscles which consist of an outer circular layer and an inner longitudinal layer. The inner surface of the body wall is lined with the simple cuboidal epithelium of the parietal peritoneum which encloses the coclon.

Body cavity or Coelom

The body cavity of the starfish is a tree body cavity or coelom, being enclosed on all sides by coelomic epithelium or peritoneum. The body cavity is quite spacious and is often referred to as the perivisceral cavity since it surrounds the viscera. The perivisceral coelom extends between the oral and aboral sides of the central disc and is continued into each of the five arms. Besides, there are other coelomic cavities such as the water vascular system, axial sinus, perihaemal canals, genital sinuses, etc. The coelom itself is filled with albuminous coelomic fluid or haemolymph. The coelomic fluid contains phagocytic coelomocytes or amoebocytes. The coelomic fluid receives soluble nutrients from the digestive system, and the metabolic wastes are transported to the dermabranchiae where the wastes pass out into the surrounding sea water by diffusion.

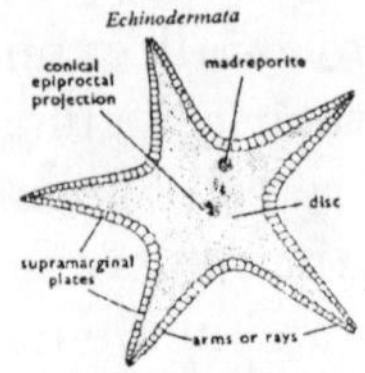

Ctenodiscus crispatus. Aboral view.

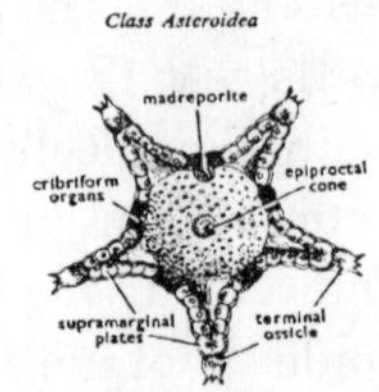

Porcellanaster caeruleus. Aboral view.

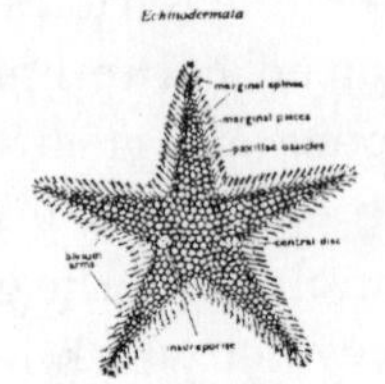

Astropecten. Aboral view.

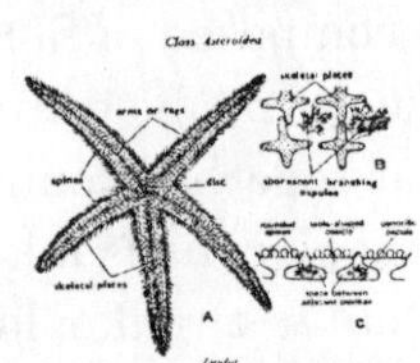

Luidia.
A - Aboral view. B - Skeletal plates and arborescent papulae at surface. C - Diagrammatic C.S. through several papillae.

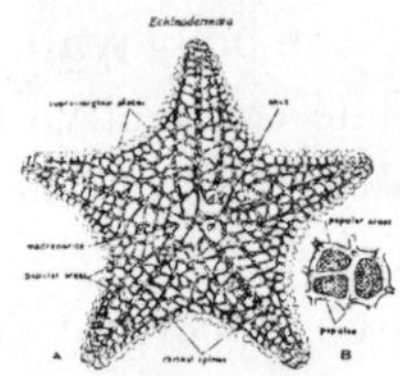

Oreaster reticulatus.
A - Entire, in aboral view.
B - Meshes of skeleton magnified to show papulae.

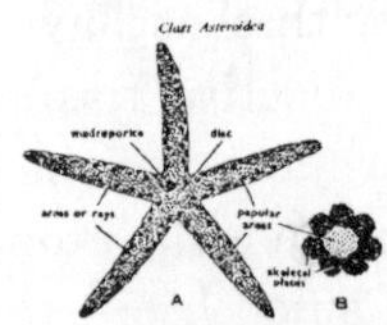

Linckia laevigata.
A - Entire in aboral view. B - A papular area enlarged.

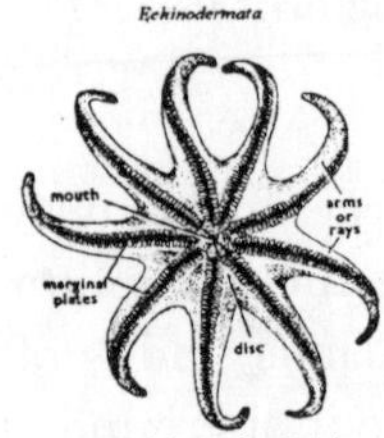

Solaster endeca. Oral view.

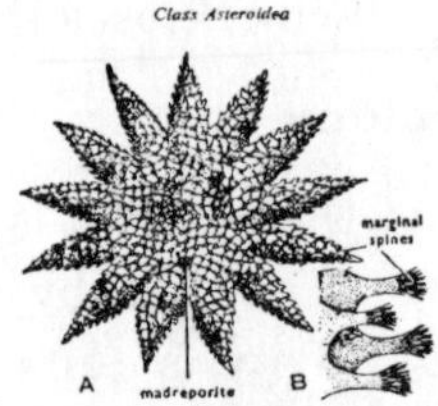

Crossaster papposus.
A - Entire in aboral view. B - Marginal spines enlarged

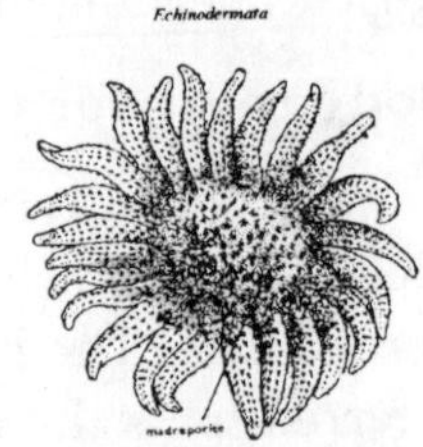

Heliaster kubiniji. entire, in aboral view.

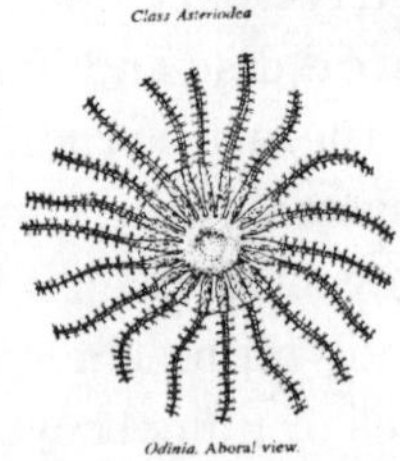

Odinia. Aboral view.

Digestive System

The alimentary canal is a straight tube, extending from the mouth on the oral side to the anus at the aboral surface. The mouth is equipped with a sphinctor muscle. It is also surrounded by peristomial ambulacral ossicles or papillae.

The mouth leads into a short oesophagus which opens above into a spacious cardiac stomach that practically occupies the greater part of the central disc. The lateral wall of the cardiac stomach is evaginated into ten lobes or pouches, two of which project for a short distance into each arm. The cardiac stomach is eversible through the mouth during feeding. The cardiac stomach is followed by the pyloric portion of the stomach. The pyloric stomach sends out within each arm, a pair of pyloric caecae (hepatic caeca) which are branched and glandular. They secrete digestive enzymes. The pyloric stomach leads into a short tubular intestine which terminates at the anus on the dorsal surface of the central disc above. The intestine diverticulates into a bilobed intestinal caecum lined with mucous and glandular cells.

The food of starfish includes worms, crustaceans, vivalve molluscs and even small fishes. The small animals are swallowed direct but the larger animals are seized by the tube feet. The secretion of the cardiac stomach and pyloric caeca are poured outside the body into the body of the prey. Intracellular digestion takes place within the stomach and pyloric caeca. Large bits of undigested food are passed out through the mouth, and only a little of it through the anus.

Water vascular system (Ambulacral system)

The water vascular system or systems of ambulacral vessel presents one of the most characteristic feature of the internal anatomy of the Echinoderms. This is a system of tubes that courses along the inner surface of the ambulacra

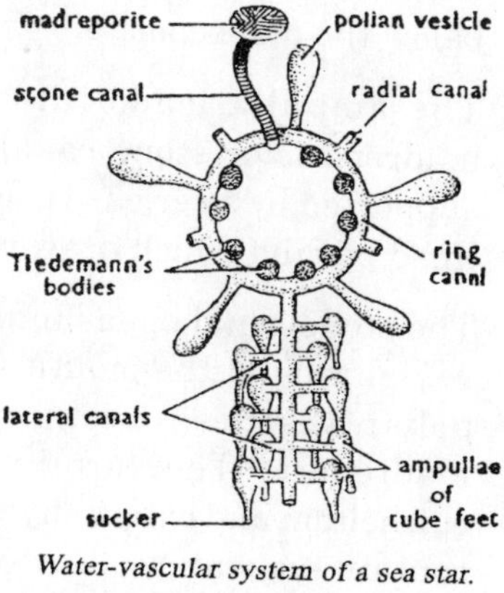

Water-vascular system of a sea star.

from the oral region and extending to their tips. The tubes are filled up with a watery fluid, and the system helps in locomotion of the animal. This system, embryologically derived from a part of the larval coelon known as hydrocoel and separated from it during the developmental stages of the animal, begins with the madreporite, a circular calcareous plate located eccentrically on the dorsal surface of the central disc.

Essentially the water vascular system of starfish consists of the madreporite, stone canal, ring canal, radial canals, Tiedmann's bodies, lateral canal and tube feet.

1. From the actinostome, a five-rayed aperture situated at the centre of the oral surface, radiate out five narrow ambulacral grooves, one along the middle of the oral surface of each arm.
2. The madreporite is situated near the anus on the aboral surface, between the bases of two arms. It is a nearly circular, flat, calcareous plate, and its surface is provided with numerous narrow radiating grooves. Large number of narrow pores are present in the grooves. The pores may be about 200 in number. Each pore leads through a pore canal in the madreporite. The pore canals unite to form collecting canals which open into an ampulla lying below the madreporite.
3. The ampulla leads into an s-shaped cylinder, the madreporic or stone canal. The walls of the canal are supported by a series of clacareous rings and its cavity is lined by cilia which draw water into the canal.
4. The stone canal opens into the ring canal, a pentagonal vessel around the mouth. The ring canal gives out four polian vesicles, one in each inter radius except the madreporic. They are lined internally with flattened epithelium and externally with coelomic epithelium. They contain sea water.

5. The ring canal gives out five radial canals, one for each arm, which runs along the arm, and lies within the arm beneath the shelter of the rafter-like ossicles of the ambulacral groove.
6. Throughout its length, each radial canal gives off two rows of lateral or podial canals which connect with the tube feet on the oral side of the ambulacral pores.
7. From the inner aspect of the ring canal arise nine small vesicles called Tiedmann's bodies or vesicles. A pair is placed by the sides of each radial canal near which madreporic opens. Tiedmann's bodies probably manufacture the amoebocytes of the water vascular system.
8. Each tube-foot is a thin-walled hollow cylinder bearing a sucker at its free outer end. A tube-foot consists of two parts, a tubular podium and a sac-like ampulla.
9. The walls of the tube-feet are muscular, and there are also valves at the junctions of tubes and ampullae. The valves regulate the flow of water. In case the tubes are extended, they come out through the ambulacral grooves. There are muscle fibres in the wall, of the ampullae.

The water vascular system plays an important role in the locomotion of the starfish. The beating of cilia of the ambulacral system causes sea water to enter through the madreporite into the stone canal. Water then passes into the ring canal and from there, it is driven into the radial ambulacral vessels. From there the water is led into the lateral branches to fill the ampullae of the tube feet. When the ampullae contract, the water is forced into tube feet. The hydrostatic pressure resulting from this insurge of fluid causes the extension or elongation of the tube feet in the general direction in which the animal is moving. The terminal discs come in contact with the substrate; the central portions of these terminal discs are withdrawn inward to form suction cups, thus rendering the tube feet firmly adherent to the substrate. Now the tube feet pivot forward on their attached terminal discs – an action which has the effect of pushing the animal forward. The longitudinal musculature in the tube feet now begins to contract and the terminal discs are released. On account of the contraction of the musculature in the tube feet, water is now forced into the ampullae. Backflow of water in the tube feet is prevented by valves present near the opening of the ampullae. When the ampullae again contract, water enters the tube feet, causing their extension and the sequence of events is repeated.

The passage of water from the exterior to the tube feet is summed up as follows:

Fluid from exterior → Madreporite
↓
Stone canal
↓
Ring canal → Radial canal → Ampullae → Tube feet
↓
Exterior

Respiratory System

Respiration is effected by gills or papulae anal also by the tube feet. The gills are contractile hollow outgrowths of the skin on the aboral surface. The gills project through small holes between the ossicles. The cavities of the gills are connected with the coelom. Cilia are present on the outer and inner surfaces of the gills. The cilia of the outer surface produce a respiratory current of water over the gills, while cilia of the inner surface cause the flow into the gills. As the coelomic fluid is exposed to the surrounding water, exchange of gases takes place. In addition to the gills, the tube feet also serve for the exchange of gases.

Nervous System

The nervous system of starfish is of a primitive type as there is no concentration of nerve cells to form a ganglion or the brain. The nerve cells with their processes and nerve fibres are scattered in a diffused state giving rise to nerve nets. According to their position the nervous system is of two types, a superficial and deep. The superficial type is ectodermal or epidermal in origin and is distributed over the epithelium of the body. The deep type originated from the coelomic epithelium, and it occurs in the coelomic epithelium of the perihaemal radial vessels. The nervous tissue of the superficial type is thickened as a band on the epithelium lining the ambulacral groove and is distinguished as the radial nerve. From this five radial nerve cords arise and they run into the arms lying within the ambulacral groove. The radial nerve also gives off nerve branches to the tube feet and they are continuous with the nervous layer of the body wall and its appendages. The deep nerve includes a central pentagon which is double and is lies aboral to the nerve ring.

Reproductive System

The sexes are separate. The gonads (testes or ovaries) are temporary structures as they are developed only during the breeding season. The gonads look like bunches of grapes. They occur as five pairs of sacs. They are interradial in position and hang from the roof of the coelomic cavity. Each gonad opens aborally by a sieve-like gonopore. Testes are grey in colour while the ovaries are orange. There is no copulatory organ or accessory gland. The gametes are discharged directly to the sea water where fertilization takes place.

LARVAL FORMS OF ECHINODERMATA

There are various types of Echinoderm larvae as the development is indirect. The various larval forms show certain common characteristics, so that it is possible to create a hypothetical larva, known as Dipleurula larva. It is believed that from this hypothetical Dipleurula larva, various other echinoderm larvae have arisen by suitable modifications in the ciliated loops and in the preoral lobe.

1. Dipleurula larva

The larva has bilateral symmetry of the body. The ventral side is concave and it bears the mouth which is surrounded by a band of cilia called the aboral band. The anus is disposed ventrally. The enlarged region in front of the mouth is known as preoral lobe that is separate from the longitudinal bands of cilia with their cross connections which form postoral loops. There are three pairs of coelomic sacs. The coelomic sac on the left becomes subdivided, and a section of it develops into the hydrocoel. The coelom communicates to the exterior. The arrangement of the ciliated loops and the relative proportion of the precoral lobe, give rise to modification of Dipleurula into different types of larvae.

2. Pentactula larva

This larva is considered to be the next evolutionary step of the Dipleurula larva. This larva possesses five tentacles around the mouth and the hydrocel is separated from the rest of the coelon so as to form the future water vascular system.

3. Bipinnaria larva

The larva of the starfish (Class Asteroidea) is known as Bipinnaria larva. It is a modification of Dipleurula. Its preoral region is elongated while the postoral region is broad. The body is encircled by two ciliated bands, of which the

Echinodermata

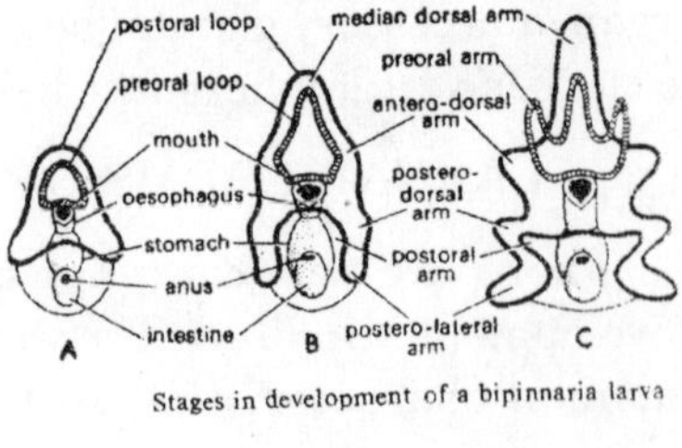

Stages in development of a bipinnaria larva

preoral is smaller, the post oral much larger. They are extended peripherally by the development of soft bilateral arms. According to their place of situation, the arms are distinguished as median dorsal, median ventral, antero dorsal and postero dorsal arms. The arms are distinctive of Bipinnaria larva. The alimentary canal is U-shaped and the mouth and anus are on the same surface. The Bipannaria is a free swimming pelagic larva. This larva swims about for several weeks, and eventually develops three brachiolar arms to become a Brachiolaria.

4. Brachiolaria larva

This type of larva is also found in Asteroidea. It is a modified stage of Bipinnaria larva. In Brachiolaria larva there are three warty processes at the anterior dorsal end and these unlike the other arms arise independently of the ciliated band. After free swimming for a few weeks, the larva settles down to the bottom of the sea and fixes itself to a substratum by means of its anterior end which becomes prolonged like a stalk. The larva then undergoes changes in symmetry. During metamorphosis the ciliated bands together with the mouth and anus disappear, a new mouth and anus are formed, the hydrocoel on the left becomes lobed surrounding the new mouth. The newly formed body becomes radially symmetrical.

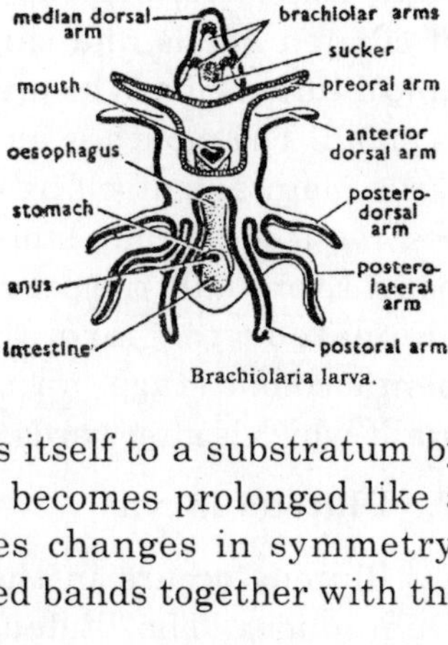

Brachiolaria larva.

5. Auricularia larva

The auricularia is the simplest larva, and it occurs in

Echinodermata

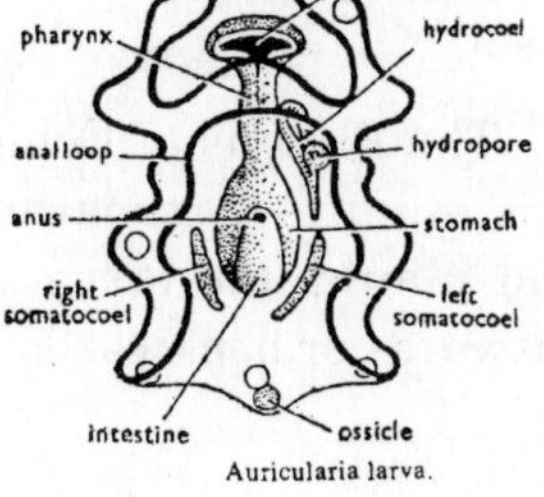

Auricularia larva.

Holothurians and Asteroids. In this larva the preoral lobe is prominent and the preoral band is large. There is a single longitudinal band of cilia which passes across the ventral surface immediately in front of the mouth and in front of the anus and bounds a depression on the ventral surface into which the mouth opens. By partial degeneration of the locomotor bands and their rearrangement, the auricularia of holothurians becomes a Doliolaria.

6. Doliolaria larva

Haeckel named the larva of Antedon (crinoid) as Doliolaria larva. It is barrel shaped with hoops of ciliated bands and carrying an apical tuft of cilia. The preoral lobe is small. The mouth is absent. The larva maintains itself by using the reserve food already stored in the body. Eventually the preoral end is elongated to form, in the pentacrinoid stage, an attaching stalk, which is afterwards absorbed.

Forms in Echinodermata

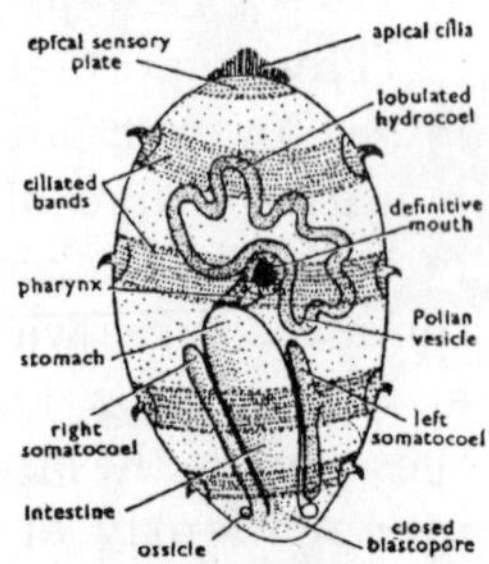

Mature doliolaria larva of *Leptosynapta inhaerens*.

7. Pluteus larva

Pluteus occurs in the life history of Echinoidea and Ophiuroidea. The Plufeus has a number of slender arms which are directed forward and supported inside by calcareous rods. The preoral lobe is small, while the postoral region becomes large. The larval body is comparable to a six-legged easel.

The Pluteus of Echinoidea is known as *Echinopluteus*. It is minute, transparent, ciliated, bilateral free swimming and planktonic larva. The arms are small and directed outwards or backwards.

The Pluteus of Ophiuroidea is called *Ophiopluteus.* This larva has fewer arms than the Echinopluteus but its postero lateral arms are the largest and are directed forwards. The arms are supported by calcareous rods.

Significance of Echinoderm larvae

The Echinoderm larvae have great phylogenetic significance. Structurally they resemble each other, and this is indicative of the fact that all the diversified Echinoderm classes might have evolved from a common ancestor.

The Echinoderm larvae have affinities with certain non-chordates and chordates. Particularly, the Auricularia and Bipinnaria larvae resemble the trochophore larva of the Annelid on one hand and Tornaria larva of Balanoglossus (Hemi-chordata) on the other. Thus, the larval forms of Echinodermata establish phylogenetic relationship of Echinodermata with some higher invertebrata like Annelida and the lower chordates like Hemichordata.

8
ACCOUNTS OF MINOR PHYLA

I. NEMERTINEA (RHYNCHOCOELA)

General Characters

The Nemertinea commonly known as the ribbon worms are found along the sea-shores under stones and among sea-weeds, a few live in fresh water and damp soil.

1. Body is elongated, cylindrical, more or less flattened ; there is no external segmentation, although internal parts often display pseudometamerism.
2. The body is bilaterally symmetrical. The anterior end has a distinct cephalic lobe which bears numerous simple eyes and sensory cells.
3. The body surface is covered with a glandular ciliated epidermis and the interior consists of muscle layers and mesenchyme.
4. The mouth is situated ventrally near the anterior end, or else the digestive tract opens by the proboscis pore so that a separate mouth is lacking.
5. Proboscis pore occurs near the anterior tip and leads into a cavity, the rhynchodaeum, to whose inner end is attached the anterior end of the usually elongated,

tubular proboscis. This lies free in a cavity, the rhynchocoel, situated mid-dorsally above the digestive tract.

6. The digestive tract is often differentiated into several regions, and the intestine frequently bears paired lateral diverticula.
7. An anus is invariably present at the posterior end.
8. The excretory system consists of one to many pairs of nephridia provided with terminal flame bulbs, but in most forms the protonephridia extend for only a small part of the body length.
9. The circulatory system consists of a pair of lateral vessels, and many forms are provided in addition with an unpaired dorsal vessel.
10. There is no special respiratory mechanism. The required oxygen for respiration diffuses through the body surface.
11. The nervous system consists of a well developed brain having four lobes, and a pair of main lateral nerves and minor nerves.
12. Sensory organs comprise of sensory nerve cells, sensory pits, eyes and cerebral organ. Sensory nerve cells are scattered in the epidermis of anterior and posterior ends. They are tactile in function. Sensory pit is a group of nerve cells found all over the body. The eyes are borne by the cephalic lobe of the body. The number of eyes varies from two to many. The cerebral organ is associated with the brain. They are inverted epidermal canals, the inner end of which is surrounded by a mass of gland and nerve cells connected with the brain. It is chemoreceptive in nature.
13. Most of the Nemertina are dioecious. Gonads consist of small sacs, each with a direct opening to the exterior.

There is no copulatory organ. Development is of spiral determinate type. The young hatches out as a free-swimming Pilidium larva.

Classification

The Nemertines are generally classified into two main groups, each of which is again sub-divided into two groups. Because of the great similarity of structures throughout the Phylum, it does not seem justifiable to consider the two main groups as meriting the rank of classes. They are, therefore, considered here as sub classes.

Sub-Class I Anopla: In this, mouth is posterior to the brain; central nervous system is situated beneath the epidermis or among the muscle layer of the body wall; proboscis is unarmed.

Order 1. *Palaeonemertini:* Body wall musculature is two or three layered; the innermost layer is circular. Dermis is gelatinous. Circulatory system has no mid-dorsal vessel, and the intestine is without lateral diverticula.

Ex. – Procarinina, Cephalothrix, etc.

Order 2. Heteronemertini: Body wall musculature is three-layers, outer longitudinal, middle circular, and inner longitudinal. Dermis is fibrous. The lateral nerve cords lie between the longitudinal and circular muscle layers. Mid-dorsal vessel is present.

Ex. – Micrura, Lineus, etc.

Sub-Class II. Enopla: Mouth is anterior to the brain. The intestine bears a caecum and lateral diverticula. Central nervous system is internal to the body wall musculature. Proboscis may be armed.

Order 1. *Hoplonemertini:* Proboscis is armed with one or more stylets. Intestine is straight with paired lateral diverticula. Mid-dorsal vessel is connected with the lateral vessels by transverse vessels.

Ex. – Tetrastemma, Prostoma, etc.

Order 2. Bdellomorpha: Proboscis is unarmed. Intestine is sinuous without diverticula. Body is provided with a posterior adhesive disk.

Ex. – Malacobdella.

Phylogenetic consideration of Nermertinea

Although in the past, Nemertines have been derived from various phyla, especially Annelida, at the present time there is little doubt of their derivation from the Platyhelminthes. Nemertines have resemblances with the Platyhelminthes in the following points:

1. Epidermis, muscle sheath, mesenchyme filling the spaces between organs.
2. Occurrence of frontal glands.
3. General plan of nervous system.
4. Structure of the ocelli.
5. Occurrence of the ciliated grooves, pits, etc., as the sense organs of the head.
6. Spiral type of cleavage.
7. Cerebral organs are regarded as derived from the auricular sense organs of flatworms.
8. Pseudometamerism is similar to that of marine triclads.

Ex.– Cerebratulus, Lineus, Prostoma, Geonemertes, etc.

The serological tests of Schepotieff (1913) in which nemertine antiserum reacted with Polyclad extract but not with annelid extract support the above conclusion of a Platyhelminth origin of the Nemertines. The Nemertines represent the culmination of the acoelomate type of construction.

Cuvier (1817) described a nemertinea under the name "Nermertes", creating a separate order. De Blainville (1828) considered the nemertines in a family "Teretularia". Ehrenberg (1828) included the nemertines under Turbellaria, order Rhabdocoela. Delle Chiaje (1844) used the term Annulosa Nemertea for Nemertines.

Oersted (1844) was the first to create a group Cestoidina that contained nothing but Nemertines. Von Siebold (1846) devised the name Nemertini for the Nemertines. Deising (1850) placed the Nemertines in a tribe Nemertinea under Turbellaria.

In 1850-1851, Max Schultze gave the correct and final account of Nemertines. We also follow him under the name Phylum Rhynchocoela or Nemertina.

After considering the above remarks of the zoologists and the many similarities of characters present by this Rhynchocoela with Platyhelminthes (as shown in the above characters), Nemertines are regarded as a class of the Platyhelminthes; but, while presenting many similarities to the flatworms, they are most highly organized in several respects, such as their possession of a circulatory system, the greater differentiation of the digestive tract and the body wall, and the presence of definite anus. They are further distinguished from all other groups of animals by the occurrence dorsal to the digestive tract of a long eversible tubular Proboscis enclosed in a hollow proboscis sheath.

It, therefore, appears justifiable to regard these Nemertines as constituting an independent Phylum.

ROTIFERA

The rotifera or wheel animalcules are microscopic free

living metazoan animals, found in abundance in fresh water, damp moss, and in the sea. The name wheel animalcule has been given to the rotifers on the ground that the rapid movements of cilia on their anterior end produce the appearance of a rotating wheel. Some of them form colonies and some form a tube round the body.

General Characters

1. The body is bilaterally symmetrical.
2. There is no internal segmentation, but there are sometimes external rings, and the attaching outgrowth or "foot" is sometimes ringed.
3. The body is distinguishable into an anterior head region, an expanded trunk and a tail-like posterior foot, which often ends in two slender toes.
4. The head region bears the mouth and a retractile ciliated disc, the corona or trochal disc with a central unciliated apical field. The corona may be surrounded by a double ciliated ring, the velum, made of an outer ciliary band or cingulum and an inner trachus. The corona or "wheel organ" is a very characteristic feature of rotifers. Corona serves locomotion, and by beating of its cilia, food particles are propelled into the mouth. Pigmented red spot or ocelli (eyes) occur near the corona.
5. The trunk may be cylindrical or variously flattened and is frequently enclosed in a lorica, often ornamented or spiny. Characteristic trunk structures of Rotifera are the dorsal and lateral antennae or palps.
6. The anus is situated at the mid dorsal line at or near the boundary of the trunk and foot.
7. The tail or foot is provided with ringed cuticle. It contains cement or pedal glands whose ducts open at the tips of the toes. The secretions of the pedal glands serve to anchor the animal during feeding.

8. The body wall consists of outmost cuticle (a selero-protein), often ringed, the epidermis – a syncytium with scattered nuclei, and a few longitudinal muscle fibres.
9. There is a spacious body cavity between the body wall and the digestive tract. It is a pseudocoelom in not being lined by a mesodermal epithelium and being regarded as equivalent to the blastocoele. The cavity is filled with fluid and a loose network formed by the union of branched amoeboid cells into a syncytium. Embryologically these cells come from the same cells that furnish the epidermis. They presumably have phagocytic and excretory function.
10. The digestive tract is lined by cilia. It consists of the mouth situated below the corona, a muscular pharynx (or the mastax which bears chitinous jaws or trophi having teeth), a short oesophagus, a sac-like stomach lined by large granular ciliated cells and supported by circular and longitudinal muscle fibres. Attached to the stomach are a pair of gastric glands. The stomach leads into a short intestine and cloaca which opens to the outside by the anus.
11. The excretory system consists of a pair of convoluted longitudinal tubules communicating by side capillary branches and the flame bulbs. The longitudinal tubules open in a vesicle or urinary bladder which contracts periodically to discharge its fluid through the cloaca.
12. The nervous system consists of a single large ganglion situated above the mastax. It gives out nerve fibres to adjacent parts.
13. The sense organs are the membranelles, styles, ciliated pits, papillae and eye spots.
14. The sexes are separate. The males are smaller than females. The male reproductive system consists of a sac-

like testis, a ciliated duct which opens into the penis or in the foot. The female reproductive system includes a single syncytial ovary and a syncytial vitellarium (yolk gland). An oviduct from the ovary opens into the cloaca.

Ex. – Floscularia, Hydatina, Mellicerata, Phylodina, etc.

PHYLUM ACANTHOCEPHALA

"The Acanthocephala are endoparasitic pseudocoelomate vermiform Bilateria without a digestive tract and with the anterior end formed of an invaginable proboscis armed with hooks" – Hyman.

Many distinctive features present by the Acanthocephala can be enumerated in the following way –

1. Body is elongated ranging from 1.5 mm. to over 2 mm. in length, flattened, and rough in life, and capable of extension and contraction.

2. The body wall consists of cuticle, syncytial epidermis with spaces, and a layer of circular muscles.

3. Beneath the body wall and the ligaments there occurs a cavity, the pseudocoel, lined by a sheath of longitudinal muscles, and filled with a fluid.

4. Neither the larva nor the adult has a digestive tract, food being absorbed directly from the intestine of the host.

5. Peculiar ligament sacs and ligament strand occur. Anteriorly the ligament sacs are attached to the posterior end of the proboscis receptacle, posteriorly they terminate on or in some part of the reproductive system.

6. Excretory system consists of two branched nephridia with cilia, and they are connected to a common excretory duct. In certain genera (Oligacanthorhynchus, Nephridiorhynchus), the flame bulbs open directly into a

sac. The number of flame bulbs in each case of protonephridia ranges from about 250 to 700. The flame bulbs are devoid of nuclei.

7. No circulatory or respiratory structures.
8. Nervous system consists of a ganglion near the proboscis and two lateral cords.
9. Sense organs are of a tactile nature. They consist of three in the proboscis and several in the male bursa and penis. In the poboscis there is a sensory organ in the centre of the tip and in some one on each side in the neck. The terminal sense organs consist of a pit.
10. Sexes are separate. The male contains a pair of testes developed in a cord-like ligament extending from the proboscis sheath to the posterior end, and the sperms pass into a duct through the posterior bell-like bursa containing the genital pore. The female has no persistent ovary. Ova develop on the ligament and are set free in the body cavity and fertilized there. The eggs develop within the maternal body into larvae. The larvae come out of the host's intestine along with the faeces of the host. The intermediate host's of the larva is generally the insect. The final host is the vertebrate like fish, birds and mammals. The adult parasites then infect the digestive tract of the vertebrate host.

Classification of Acanthocephala

Following Meyer (1930) and Van Cleave (1936), the Acanthocephala are divided into three orders. Later, Van Cleave (1948) arranged the Phylum into two classes each with two orders. Hyman is unable to accept this arrangement, for the differences among the Acanthocephala are not sufficient to warrant to erection of classes and the sole difference between two of the orders is the presence of or absence of body spines.

Order 1 Archiacanthocephala: Main lacunar channels are median; proboscis spines are arranged concentrically; with protonephridia; females with two persistent ligament sacs; males with 8 cement glands; mostly in terrestrial hosts.

Order 2 Palaeacanthocephala: Main lacunar channels are lateral; proboscis hooks are arranged in alternating radial rows; without protonephridia; ligament sacs ephemeral in females; males with mostly six cement glands; mostly in aquatic hosts.

Order 3 Eocanthocephala: Main lacunar channels are median; proboscis hooks are arranged radially; without protonephridia; ligament sacs persistent in females; cement glands syncytial with a reservoir; in aquatic hosts.

Phylogenetic Consideration of Acanthocephala

The acanthocephalan worms were noticed about the beginning of the 18th century but were not clearly distinguished from other intestinal worms until 1771 when Koelreuther proposed the name Acanthocephalus for one from a fish. In 1776, Zoega and OF. Muller independently gave the name Echinorhynchus to a similar fish parasite. Later, Muller described a number of species under the name Echinorhynchus, and this name came into general use for the type of worm in question. Zeder (1803) gave these worms a common name "Hakenwurmer" (hooked worms), and Rudolpi (1809) transcribed this into the form Acanthocephala. During the 19th century many species were described under the one generic name "Echinorhynchus", and it was not until after 1890 that discriminating taxonomic work began to be done on the group, resulting the definition of numerous genera and families.

At present two prominent views have been asserted for the association of the Acanthocephala with the Platyhelminthes or with the Aschelminthes.

Affinities with Platyhelminthes

Similarities with Cestoda, Trematodes and Turbellaria

1. An armed proboscis invaginable or withdrawable into the interior occurs in Cestodes as in the Acanthocephala.
2. The cuticle of the Trematodes is similar to that of the acantho- cephalans.
3. Possession of a syncytial nucleated epidermis in the acanthocephalans has brought similarity with Turbellarians; and added to this, the epidermis with its strinted appearance has suggested closer resemblance with the latter than do with the Aschelminthes.
4. The body wall musculature of circular and longitudinal layers of the acanthocephalans is distinctly Platyhelminthic in pattern.
5. Flame bulb (protonephridia) of the Acanthocephala corresponds with that of the Platyhelminthes.
6. Reproductive system of the acanthocephalans is very much suggestive of flatworms. If the cement glands be taken to correspond to prostatic glands, then the general picture in the male of paired testes and sperm ducts, penis papilla and muscular bursa recalls conditions found in flatworms.
7. Embryology of the acanthocephalas is strongly suggestive of that of Cestodes, viz, (i) in the enclosure of a central mass by surface cells; (ii) early separation of the epidermis from the anterior mass; (iii) the formation of the cells; (iv) the eventual production of a larva showing the differentiation except for the six hooks at one end.

Dissimilarities: Dissimilarities are many.

(i) Division of the body into prosoma and trunk absent in Flatworms.

(ii) No superficial segmentation in Flatworms.

(iii) No digestive tract in Acanthocephala, whereas it is characterised by its branched type in most of the Platyhelminthes.

(iv) Flame bulb devoid of nuclei in Acanthocephala.

(v) In Platyhelminthes, mesoderm is present between ectoderm and endoderm.

(vi) Platyhelminthes are acoelomate whereas the Acanthocephala are pesudoelomate.

(vii) A close relationship of nephridial and reproductive systems that is present in Acanthocephala, is however, wanting in Flatworms.

(viii) Sexes united in Platyhelminthes whereas sexes are separate in Acanthocephala.

Affinities with Aschelminthes

Similarities

(i) Acanthocephalic division of the body is found in Aschelminthes, for example, Prioapulids, Gordiacean larva.

(ii) Armed proboscis of Aschelminthes corresponds to that of the proboscis of Acanthocephala.

(iii) Superficial segmentation of Acanthocephala is like that of Aschelminthes.

(iv) Cuticle of the acanthocephalans is practically identical chemically with that of the Nematodes.

(v) Epidermis is synchytial in both the cases.

(vi) Division of the pseudocoel by partition and tissues resembling mesenteries is distinctly aschelminthic.

(vii) The ligament strand representing the greatly reduced digestive tract of the Acanthocephala is similar to that of male rotifers.

(viii) Similarity of the flame bulb and protonephridia in between the two.

(ix) A close relationship of nephridial and reproductive system in Acanthocephala is exemplified among the Aschelminthes; in rotifers too both systems enter a common cloaca which according to Haffner's view is represented in Acanthocephala by the bursa.

Dissimilarities between the two are little.

(i) No proboscis in Platyhelminthic Nematode.

(ii) The cuticle in some of the Aschelminthes is subject to molt during growth.

(iii) No circular muscles in Platyhelminthic Nematode.

(iv) Presence of straight or curved digestive tract with pharynx in Aschelminthes.

(v) Developmental cleavage is of determinate type in Aschelminthes.

Conclusion

Though there is dissimilarity of characters between Acanthocephala and Aschelminthes, the above analysis indicates closer affinity with Echinoderids, Priapulids and Gordiods. And hence, in their general structural plan, that of the pseudocoelomate grade, the acanthocephalans certainly more nearly resemble the Aschelminthes than they do with the Platyhelminthes.

The position in the animal kingdom of the Acanthocephala is very much uncertain. The foregoing comparisons do not furnish a tentative answer to the question whether the Acanthocephala are allied to the

Platyhelminthes or to the Aschelminthes. No doubt, the general structure to assert its affinity among the animal kingdom is rather on the Aschilminthic side, whereas the embryology presents more points of resemblance with the Platyhelminthes. Chitwood (1940) and Van Cleave (1941) favour a Platyhelminthic affinity, and on this basis the former has proposed a super Phylum Parenchymata to embrach the Flatworms, Nemertines and acanthocephalans.

Prof. Cuvier too wanted to put together in a group of parasitic worms. In 1851, Vogt first distinguished the Flatworms from the roundworms under the name Plathyelmia and Nematelmia. And in the group Nematelmia, he included Gregarines, Nematodes, acanthocephalans and Gordiaceans. In 1859, Gegenbaur modified as Nemathelminthes in which Nematoda, Acanthocephala and Gordiucea were included. Schneider (1866) added to the assemblage the Chaetognatha. And lastly in the Kukenthel Krumbach Handbuch der Zoologic, the Phylum Nemathelminthes is conceived as consisting of six groups. Rotifera, Gastrotricha, Kinorhyncha, Acanthocephala, Nematoda and Nematomorpha.

PHYLUM ECTOPROCTA (BRYOZOA)

Characters

1. The Ectoprocta (Bryozoa) are sedentary and mostly colonial animals, but the erect forms may have a calcareous ectocyst and may attain flexibility in which case there are horny joints at intervals.

2. Each individual body is divisible into two parts: (i) the zooecium or body wall, alimentary canal, the tentacles and tentacular sheath into which the tentacles are withdrawn when contracted. The polypide also carries the semicircular ciliated ring of the lophophore which is provided with tentacles.

3. The body wall consists of cuticle, ectoderm and a delicate layer of cells which lines the body cavity. The peritoneal

cells are ciliated in the Phylactolaemata in which group there is also a thin muscular layer in the body wall. Special muscles of the body include the longitudinal muscles, retractor muscles and the parieto-vaginal muscles.

4. The body cavity is a true coelom, filled with a colourless fluid and colourless corpuscles. It is lined by coelomic epithelium and divided into two chambers viz., the general body cavity and the ring canal of the lophophore which is prolonged into the tentacles.

5. Mouth is at the anterior end of the body in the midst of the circlet of ciliated tentacles and the disc bearing it and the tentacles is called the lophophore, which is either circular or is drawn out into two lobes, and the tentacles are set along its edge. The alimentary canal is bent on itself, and it consists of (i) an anterior elongated ciliated oesophagus often dilated at its upper end into a muscular pharynx, (ii) a spacious stomach with a blind backward prolongation, the hind end of which is attached to the body wall of the funiculus (tissue strand) and (iii) a narrow intestine which is bent nearly parallel to the pharynx and is directed upwards. The intestine opens by the anus near the mouth outside tentacular crown.

6. There is no definite excretory system. The leucocytes or the corpuscles of the coelomic fluid collect the nitrogenous waste matters. However, in Phylactolaemata, there exists a pair of nephridia for excretion. The two nephridia open into an excretory vesicle which again opens outside by a minute pore above the anus.

7. Vascular system and special sense organs are absent.

8. The nervous system consists of a ganglion placed on the oesophagus between the mouth and the anus. In the Phylactolaemata the ganglion is placed in the concavity of the lophophore, and is attached to the oesophagus by

a delicate circum-oesophageal ring. It gives rise to numerous nerve branches to the tentacles and the oesophagus. According to Fr. Muller, there is in Serialaria a so called colonial nervous system which connects the individual zooids of a colony.

9. The zooids are usually hermaphrodite. Testes develop either on the upper part of the funiculus, or near the point of attachment of the latter to the body wall. Ovaries are placed on the body wall near the anterior end of the zooecium.

10. Reproduction is partly sexual and partly asexual. Fertilization occurs within the zooecium. The eggs are sometimes laid and developed in the sea. The egg develops into a kind of trochophore with ciliated bands.

Classification

Ectoprocta is divided into two classes:

Class 1. *Phylactolaemata*

In them the lophophore is horse-shoe shaped. The mouth is overhung by a lip (epistome).

Ex. – Plumatella, Cristatella, etc.

Class 2. Gymnolaemata

In these the lophophore is circular. The mouth is not provided with an epistome.

Ex. – Bugula, Crisia, Flustra, etc.

PHYLUM BRACHIOPODA

The Brachiopoda are solitary, fixed apparently unsegmented coelomate Bilateria. They are found in shallow or deep waters.

1. The body is enclosed within a bivalve shell, and in the

presence of this type of shell, the Brachiopods resemble superficially the vivalve mollusc (or the mussels). However, the shells of Brachiopods are placed dorsally and ventrally and so are different from those of the mussels in which they are on the two sides of the body. The posterior end of the body is extended into a stout muscular stalk or peduncle for attachment of the body to a submerged rock or stone. The posterior end of the shell narrows, and the ventral valve projects behind the dorsal and may be produced into a short beak that is perforated to allow the stalk to protrude. The dorsal shell often bears on its internal surface a calcareous shelly ribbon or loop, which projects freely into the cavity enclosed between the two valves.

2. Lying below the shells and covering the soft parts of the body are too large mantles which are formed by reduplication of the body wall. Between the dorsal and ventral lobes of the mantle is a wedge-shaped space, the mantle cavity.

3. Lying within the mantle cavity is a large horse-shoe shaped ciliated structure called the lophophore. It is curved and it carries a row of tentacles. A ciliated groove or gutter runs on the coiled limbs of the lophophore. The lophophore contains prolongation of the coelom into which project the digestive glands.

4. The body wall consists of an outer layer of epidermis, a larger connective tissue, and a ciliated coelomic epithelium lining the body cavity. The musculature consists of abductor muscles and adjustor muscles which help to close and open the shell.

5. The body cavity is a true coelom. It extends into the folds of the mantle. The coelom is divided into a right and left half by a dorso-ventral mesentery. It is communicated with the exterior by one pair of nephridia. The coelom is

lined by ciliated coelomic epithelium and is filled with fluid containing colourless corpuscles.

6. The mouth is situated in the middle of the lophophore in a transversely directed buccal groove. The mouth leads into a ciliated alimentary canal, which is V-shaped, comprising a sac-like stomach and intestine. The intestine opens outside by the anus. A digestive gland opens into the stomach.
7. Excretory organs consist of one or two pairs of nephridia. Each nephridium opens in the body cavity by a funnel-shaped nephrostome and externally on either side of the mouth.
8. The circulatory system consists of a longitudinal vessel in the dorsal mesentery. One region of this vessel is contractile and is distinguished as the heart which is placed on the dorsal side of the stomach. Blood vessels arising from the heart supply the tentacular region, oesophagus and the generative organs.
9. The nervous system comprises of a supra-oesophageal ganglion in front of the mouth and a larger sub-oesophageal ganglion behind the mouth. The two ganglia are connected by circum-oesophageal connectives. Nerves arising from the sub-oesophageal ganglion innervate the mantle lobe, the arms, abductor muscles and the peduncle.
10. The sexes are separate. But some may be hermaphrodite. The gonads look like thick yellow bands or ridges, projecting into the body cavity. Ripe sex cells are released into the body cavity from where they pass to the exterior through the nephridia.

Classification

The Phylum Brachiopoda is divided into two classes.

Class 1. *Inarticulata or Ecardines:* The valves of the shell are alike, but without hinge, being held together by muscles. There is no shelly loop; anus is present.

Ex. – Crania, Lingula, etc.

Class 2. *Articulata or Testicardine:* Valves are dissimilar; a hinge is present with a beak for the peduncle and shelly loop for the lophophore; no anus.

Ex. – Rhynchotrmata, Magellania, etc.

Phylogenetic consideration of Brachiopoda

Brachiopods resemble superficially the bivalve molluscs in having bivalve shell. But the Brachiopods differ from the molluscs in the position of the valves, the peduncle, the arms and in the interior organization of the body.

Brachiopods resemble the Chaetognatha in the formation of the coelom and mesoderm which is enterocoelic in both. However, the Brachiopods differ from Chaetognatha in the organization of the adult structures.

The larva of Brachiopoda is suggested to be of trochophore type. Presence of setae secreted in cutaneous sacs, indication of segmentation plan of the nervous system and relations of the coelom in the adult also suggest annelidan affinities.

The arms with the tentacles of Brachiopods are comparable to the lophophore and the lip to the epistome of Phoronis. The caudal region of the Brachiopods is considered to be equivalent to the evaginated foot of the Phornis.

PHORONIDA

Phoronids are cylindrical, elongated worms to be met with in all the chief regions of the sea at moderate dephts. It was first discovered by Dr. S. Wright (1856) and until now only six species are known. Each worm is enclosed in a leathery tube within which it is capable of being completely retracted.

A number of individuals may live in the same tube, but never connected with each other.

General Characters

1. The length of the body may be upto 127 mm. which is totally unsegmented.
2. At the anterior end there is a crown of numerous, slender, ciliated tentacles borne on a horse-shoe shaped lophophore the lateral cornua of which are spirally coiled in larger species. These are supported by mesodermal skeleton and are retractile.
3. Between two rows of tentacles carried on by the lophophore there is a groove leading into the median mouth.
4. A lateral flap of body wall, the epistome (a broad lobe) overhangs the mouth.
5. The anus is dorsal and it lies outside the lophophore at the summit of a median longitudinal ridge. On each side of the anus is the external opening of the nephridia.
6. The body wall comprises of the cuticle, an ectodermal epidermis, followed by a basement membrane. Below the basement membrane lies the musculature consisting of circular and longitudinal layers. The last internal layer is the parietal layer of coelomic epithelium.
7. The coelom is distinct and is lined by coelomic epithelium. It is divided at the level of the lophophore by a transverse septum into (i) a small anterior section continued into the epistome and the tentacles, and (ii) a large posterior portion, subdivided by longitudinal mesenteries.
8. The alimentary canal is bent on itself to form a loop (U-shaped) as in Bryozoa. It consists of the oesophagus, the

stomach, the intestine and the rectum which opens at the anus. The wall of the stomach is glandular.

9. The excretory organ consists of two nephridia which also act as generative ducts and open externally on each side of the anus. The excretory organ of Actinotracha larva is provided with solenocytes.

10. The vascular system is a closed tube with two main longitudinal trunks, which are contractile and continuous aborally but giving off branches to the gut wall and gonads separately; anteriorly they pass through the septum, and are both connected with the single vessel found in each tentacle.

    Blood consists of colourless plasma containing nucleated red corpuscles (haemoglobin) in suspension.

11. The nervous system lies in the skin immediately below the epidermis (and outside the basement membrane). There is a special concentration of it round the mouth in the form of a circumoral ring which is thickened into a ganglion between the mouth and the anus, which is known as supra-oesophageal ganglia. Nerves arising from the ring innervate the tentacles, nephridia, etc. There are two tubes lying along the insertion of the lateral mesenteries. These are probably large nerve fibres comparable to the giant fibres of earthworms.

12. There are no special sense organs. However, there are two ciliated pits on the dorsal side of the base of the inner series of the tentacles. These pits are either sensory or glandular.

13. Phoronids are hermaphrodite. The gonads lie on the left side of the stomach in the aboral region of the body cavity. Ova and sperms are developed in the coelom towards the posterior end from the cells on one of the large blood vessels. When mature these pass out through the

nephridia to the spaces enclosed by the tentacles where the ova are fertilized.

14. The eggs are small and zygote undergoes first part of its development being entagled in the tentacles of the mother and hatch out as a free swimming larva, the Actinotracha larva, which after a period of free life undergoes a remarkable metamorphosis to become an adult.

Affinity

Placing properly the group phoronida among the various Invertebrates is a grave biological oddity, and gave rise to considerable divergence of opinions among the authorities. Naturalists like W.H. Caldwel, A.T. Masterman and others have shown its affinities with different animals as a result of which the position of phoronida still remains a disputed question.

I. Affinities with Bryozoa (by Caldwell, 1883-1885)

Similarities

1. The dorsal surface of the body is reduced.
2. The space between the mouth and anus is very short.
3. The precoral lobe of the larva almost entirely disappeared in the adult.
4. Presence of horse-shoe shaped lophophore.
5. Presence of epistome.
6. Presence of a pair of very similar nephridia.
7. Body cavity is divided very unequally by a septum.
8. The alimentary canal is bent on itself to form a loop.

Dissimilarities

In Phoronis, line between the mouth and anus is closed while the anus is posterior, outside the lophophore in Bryozoa.

Remark

Of the resemblances between the two groups, the lophophore and epistome are not so important while on the other hand they are doubtful points of anatomy.

II. Affinities with Brachiopoda

Similarities

1. Dorsal surface is reduced.
2. The space between the mouth and anus is short.
3. Preoval lobe of the larva almostly or entirely disappear in the adult.
4. Presence of lophophore.
5. Lip of Brachiopoda is comparable to epistome of Phoronis.

Dissimilarities

1. Flexure of intestine does not agree.
2. The anus is dorsal in Phoronis, ventral in Brachiopoda.
3. General shape differs.

Remark

In our present state of scanty knowledge regarding the development and real position of anus in the Brachiopoda too much weight must not be added to these objections. The possible presence of three mesoblastic segments and the relation of nephridia and coelom are very characteristic. So it may be said that Brachiopoda is allied to Phoronida and at the same time a highly speculative one.

Affinities with Annelida

The following are the similarities between the Trochophore and Actinotrocha larva.

1. Presence of ciliated ring, apical plate carrying cilia and presence of eye spot.

2. Excretory organs of Actinotrocha are comparable to that of Trochophore.

Remark

In view of differences in many points between the two and because Trochophore larva itself may be found in many groups, too much attention must not be paid to it as a clue to phylogenetic relationship.

Affinities with Enteropneusta

Similarities

1. Two glandular pockets opening into the anterior end of stomach in Phoronis comparable to the notochord of Enteropneusta.
2. Presence of 3 sections of body cavity in both.
3. Ectodermal pit of Phoronis on ventral side of preoral lobe is comparable to the sub-neural gland of Enteropneusta.

Dissimilarities

1. Gastric pouches not homologous with notochord.
2. Coelom of Phoronis is probably haemocoel.
3. Unpaired coelomic cavity in Actinotrocha but paired in Tornaria larva of Enteropneusta.

Conclusion

The phylogenetic affinities of Phoronida have long been a matter for speculation. At one time, the Phoronids have been linked with the annelids, gephyreans, ascidians, brachiopods and ectoprocts. Masterman (1900) attempted to relate the Phoronids to the Hemichordata. Recently there has been a tendency to consider the Phoronida as an independent phylum.

SIPUNCULIDA

The Sipunculids were formerly grouped with Priapulida

and Echiurida under the phylum Gephyrea, but it is improbable that the thre are nearly related. Later, they were separated because of the great differences among them.

Sipunculids are widespread marine animals. They are sedentary, burrowing in the sand or mud. Some of them excavate the burrows or live on rock or oral crevices.

Characters

1. They are elongated worms of yellowish or greyish colour. The body is divisible into an anterior narrow introvert and a large posterior trunk. The introvert ends in an anterior oral disc, bordered by tentacles of variable number, size and complexity. A smooth zone, the collar may be found immediately behind the tentacles. Minute spines or hooks and sensory or glandular papillae stud the surface of the introvert. The introvert can be retracted into the trunk. The trunk has no hooks or spines, but usually has sensory and glandular papillae. The mouth is situated at the interior end of the introvert while the anus is on the dorsal surface about one-third of the length of the body from the anterior end.

2. The body wall is composed of layers of cuticle, epidermis, dermis, circular muscle, longitudinal muscle, and peritoneum. The layers vary in thickness and complexity with the size and habits of the species. Introvert retractor muscles merge with the longitudinal muscle layers of the body wall and insert at the anterior tip of the introvert, on the upper end of the oesophagus.

3. The coelom is spacious and undivided; only vestiges of mesenteries are present. The coelom is filled with the coelomic fluid in which float several types of corpuscles.

4. The alimentary canal is U-shaped, thin walled tube. It begins at the mouth which opens into a short pharynx, or leads directly into the slender oesophagus. The

ascending arm of the gut is a coiled intestine which straightens and continues as the rectum. The rectum opens at the anus, situated on the dorsal side of the anterior region.

5. The excretory organ consists of 1 or 2 nephridia, often called brown tubes, opening externally on the ventral surface in the anterior region near the anus.
6. The circulatory system consists of dorsal and ventral vessels with ring sinus below the tentacles.
7. The nervous system is like that of Annelida, but the only ganglia are the cerebral ganglia, and the lateral nerves show no sign of a metameric distribution. Nerves to the retractor muscles usually arise from the circumenteric ring.
8. Sense organs consist of scattered neuro-sensory cells, papillate sensory buds, and protrusible ciliated pits. A ciliated region, the nuchal organ, on the dorsal edge of the oral disc is probably a chemoreceptor.
9. Sexes are separate. The gonads are peritoneal outgrowths, located at the junctions of the retractor muscles and the body wall. The germ cells are released in the body cavity where they mature. Fertilization is external. The larva is usually similar to a trochophore.

   Ex. – Sipunculus, Dendrostomum, etc.

ECHIURIDA

The Echiurida are marine worms, inhabiting the mud and sand between rocks under shallow coastal waters of all warm and temperate seas. They make up a small phylum of Annelid like.

Characters

1. The body consists of a sausage-shaped – cylindrical trunk

and an extensible grooved poboscis, corresponding to the annelid prostomium. The proboscis is extremely sensitive to touch and is probably a centre for chemoreception. It secretes a mucous cover on which organisms and small particles stick, to be conveyed to the mouth by a ciliated groove.

2. The body wall consists of an epidermis, a dermis and layers of longitudinal and circular muscles, followed by the peritoneum. The body surface has several types of glandular cells and sensory papillae. A pair of large, hooked, ventral setae project from setal sacs. They are secreted in the same manner as the setae of annelids.
3. The body cavity is spacious and undivided. The coelomic fluid contains amoebocytes and haemocytes containing haemoglobin.
4. The digestive tract is greatly coiled. The mouth lies at the base of the proboscis. The mouth opens into a short buccal tube, leading to a large pharynx. The oesophagus is a narrow tube leading to the midgut. Between the oesophagus and intestine lie a muscular gizzard and stomach. The intestine straightens and continues as a short rectum, from which a pair of diverticula the rectal sacs, protrude. The anus is terminal. A narrow ciliated tube, the siphon, parallels the intestine for some distance and then rejoins it.
5. Excretory organ includes one to many nephridia, with nephrostomes and nephridiopores.
6. The circulatory system consists of dorsal and ventral vessels which are connected anteriorly. The ventral vessel runs above the nerve cord. Blood passes to the dorsal vessel through contractile circum intestinal and circum oesophageal vessels. The dorsal vessel arises at the junction of the most posterior circum intestinal vessels

and runs forward to the anterior tip of the proboscis. Here it divides to form right and left branches, which pass along the proboscis margins and join together below the mouth to form the ventral vessel.

7. The nervous system consists of a ventral nerve cord lying within the body wall. This nerve cord divides in front into two which pass round the oesophagus and extend upto the anterior end of the proboscis, where they join one another.

8. Sexes are separate. Reproductive elements are formed on the walls of the body cavity, and are shed into it. After they have ripened, they escape through the nephridia. Development is accompanied with a trochophore larva.

   Ex. – Bonellia, Echiurus, etc.

Phylogenetic consideration

Similarities in early embryonic development, as well as the larval construction and many features of the adult organism, provide good evidence of a close link between Echiurida, Sipunculida and Annelida, although sufficient differences to justify recognizing them as separate phyla exist.

Similarities with Sipunculida

1. Absence of body segmentation in the adult stage.
2. The nephridia of Echiurida are comparable to brown tubes of Sipunculida.
3. There is a single nerve cord in both.

Similarities with Annelida

1. Possession of setae and their mode of formation.
2. Segmentation is well marked in the larval stage, although it is feeble in adult.

3. Nephridia are segmental.
4. Like Bonellia, Sternapsis a genus chaetopoda also possess a bifid proboscis.

QUESTIONS AND ANSWERS

PHYLUM PROTOZOA

Q.1. Classification of Protozoa is based on —

(A) Food vacuole, (B) Cyst structure,

(C) Types of locomotory organelles, (D) Habit and habitat.

Ans. (C)

Q.2. How Protozoa is considered immortal ?

Ans. *Amoeba* is considered immortal as natural death does not occur in *Amoeba*.

Q.3. Amoebic dysentery or Amoebiasis is caused by

(A) *Entamoeba gingialis,* (B) *E. Coli,*

(C) *E. hystolytica.* (D) All the species of Entamoeba

Ans. *(C) Entamoeba hystolytica.*

Q.4. *Entamoeba histolytica* lives in the

(A) Pus pockets of pyrrhoea. (B) Stratum corneum.

(C) Muscular tissue. (D) Intestine and colon.

Ans. (D) Intestine and colon.

Q.5. *Entamoeba gingivalis* is commonly known as

(A) Mouth amoeba, (B) Trophozoite,

(C) Amoebic hepatitis, (D) None of these.

Ans. (A) Mouth amoeba.

Q. 6. *Paramecium caudatum* has—

(A) Single nucleus, (B) Two micronuclei of the same size,

(C) One micronucleus and one meganucleus,

Ans. (C)

Q. 7. *Trypanosoma gambiense* lives in human body in

(A) Lymph, (B) Blood,

(C) Cerebro spinal fluid, (D) Blood and cerebro spinal fluid.

Ans. (D)

Q.8. *Trypanosoma gambiense* is transmitted by—

(A) House fly. (B) Mayfly.

(C) Sand fly. (D) Tse-Tse fly.

Ans. (D)

Q.9. Sleeping sickness is caused by —

(A) *Trypanosoma.* (B) *Entamoeba*

(C) *Plasmodium,* (D) *Leishmania*

Ans. (A)

Q.10. At what stage does a sleeping sickness patient develop Gambia fever ?

Ans. While the parasites are in the blood, the infected man develops Gambian fever.

Q.11. Name a Protozoa that possesses pellicle. What is the structure of the pellicle? What is the use of pellicle in that animal.

Ans. *Trypanosoma gambiense* is a Protozoa that possesses pellicle. Pellicle is a thin, elastic and firm covering the body of the animal. As it is made of fine fibrils which run along the whole length of the body. These fibrils are called microtubules. Microtubules of the pellicle help maintain the shape of the organism while it swims in the blood plasma.

Q. 12. The order in which Leishmania belongs is

(A) Diplomanadi. (B) Kinetoplasti.

(C) Hypermasti. (D) Rhizomastigi.

Ans. (B)

Q. 13. The life cycle of malarial parasite in liver is called—

(A) Erythrocytic stage. (B) Pre-erythrocytic stage.

(C) Exo-erythrocytic stage. (D) Post-erythrocytic stage.

Ans. (B)

Q. 14. Which stage of *Plasmodium* is injected in the blood when an infected anopheles mosquito bites a man ?

Ans. Sporozoite.

Q. 15. Name the hosts of *Plasmodium vivax.*

Ans. Man (primary host) and Anopheles mosquito (Secondary or intermediate host).

Q. 16. The exo-erythrocytic schizogony of the Plasmodium occurs in the —

(A) Liver cells of man. (B) Red blood cells of man.

(C) Intestine of man. (D) Stomach of the mosquito.

Ans. (A)

Q. 17. To which of the following classes will you place Plasmodium.

(A) Ciliata. (B) Sporozoa.

(C) Mastigophora. (D) Rhizopoda.

Ans. (B)

Q.18. Which of the following belongs to the class Matigophora.

(A) Noctiluca. (B) Paramecium.

(C) Mocystis. (D) Amoeba.

Ans. (A)

Q. 19. Asexual reproduction in Paramecium is brought about by —

(A) Simple binary fission. (B) Multiple fission.

(C) Sporulation. (D) transverse binary fission.

Ans. (D)

METOZOA

Q.1. What are metozoa ?

Ans. Metazoa may be defined as holozonic multi-cellular organisms which develop from embryos. Their body cells are generally differentiated into tissues and organs which are specialized for different functions.

Q. 2. What are lower and higher metozoa ?

Ans. Porifera, Coelenterata, Platyhelminthes, Aschelminthes, Ectoprocta and Acanthocephala are generally called lower metazoa. They are unsegmented, radially or bilaterally symmetrical, diploblastic or triploblastic and acoelomate or pseudocoelomate.

Mollusca, Annelida, Arthropods, Echinodermata and Chordata are known as higher metazoa. They are triploblastic and truly coelomate animals.

Q.3. What is metamerism ? Give the various theories of origin and evolution of metamerism in Metazoa.

Ans. See pages.

PORIFERA

Q.1. The classification of Porifera is based chiefly on types of —

(A) Skeleton. (B) Cleavages.

(C) Reproduction. (D) None of these.

Ans. (A)

Q.2. Most of the sponges are found in —

(A) Fresh water. (B) Warmer sea water.

(C) Brakish water. (D) Turbid water.

Ans. (B)

Q.3. Sponges are sedentary in nature. What helps in their dispersal ?

(A) Locomotion of the adult animal. (B) Their egg.

(C) Their free floating larva. (D) Gemmules.

Ans. (C)

Q. 4. In sponges a constant current of water passes through ostia into spongocoel and comes out through Oscula. This is called—

(A) Water vascular system. (B) Feeding current.

(C) Canal system. (D) Circulating current.

Ans. (C)

Q.5. What is the function of Choanocyte cells ?

Ans. Choanocyte cells are used in feeding and ensuring the flow of water with inside the Spongocoel.

Q.6. In a museum three specimens of sponges are displayed. The first specimen is vase-shaped, branched with terminal oscula. It possesses calcareous spicules and large choanocytes.

The second specimen is large having siliceous spicules, composed of six rays. The canal system is leuconoid type and possesses small choanocyts.

The third is massive and brightly coloured and have no skeleton. This possesses canal system of leuconoid type.

Name the Class of these three different sponges.

Ans. First Specimen — Class Calcarea

Second Specimen — Class Hexactinellida

Third Specimen — Class Demospongiae

Q. 7. Mention the three types of canal system met within sponges.

Ans. 1. Asconoid type.

2. Syconoid type

3. Leuconoid type.

Q.8. Incurrent canal communicates with the radial canal through —

(A) Apopyle. (B) Prosopyle.

(C) Dermal ostia. (D) None of these.

Ans. (C).

Q.9. Oscula are meant to

(A) Admit a water current. (B) Pass out water current.

(C) Digest food. (D) Produce new ostia.

Ans. (B)

Q. 10. Central cavity in a sponge is

(A) Coelom. (B) Spongocoel.

(C) Osculum. (D) Ostia.

Ans. B

Q.11. In sycon the adjacent incurrent and flagellate canals are communicated through—

(A) Prosopyles. (B) Apopyle.

(C) Gastric ostium. (D) Paragastric cavity.

Ans. (A)

Q.12. What is the name given to the excrement pore of sponges ?

Ans. Apopyle.

Q.13. Amoebocytes which secrete spicules are—

(A) Chromocytes. (B) Thesocytes.

(C) Scleroblasts. (D) Spongioblasts.

Ans. (C)

Q.14. In sycon, the choanocytes line the—

(A) Spongocoel, *or* Paragastric cavity.

(B) Incurrent canal, (C) Radial canal.

(D) Excurrent canal.

Ans. (C)

Q.15. The choanocytes or flagellated cells are—

(A) Ectodermal in origin. (B) Endodermal in origin.

(c) Mesodermal in origin. (D) Cuticular in origin.

Ans. (B)

Q.16. Mesoglea is secreted by—

(A) Coenosarc.

(B) Endosarc.

(C) Ectoderm and endoderm.

(D) Gastrovascular cavity.

Ans. (C)

Q.17. Amoebocytes in sponges occur in the —

(A) Epidermis. (B) Choanocytes.

(C) Endodermis. (D) Mesenchyme.

Ans. (D)

Q.18. Spicules are secreted by —

(A) Scleroblasts. (B) Choanocytes.

(C) Pinacocytes. (D) Porocytes.

Ans. (A)

Q.19. Sponges are —

(A) Triploblastic. (B) Monoblastic.

(C) Diploblastic. (D) None of the above.

Ans. (C)

Q.20. Digestion in sponges is —

(A) Intracellular. (B) Intercellular.

(C) Both of the above. (D) None of the above.

Ans. (A).

Q.21. The cells participating into reproduction of sponges are —

(A) Pinacocytes. (B) Choanocytes.

(C) Archaeocytes. (D) Porocytes.

Ans. (C)

COELENTERATA

Q.1. What are the nutritive zooids of Obelia ?

Ans. Polyps or hydranths.

Q.2. Give one point of similarity between Polyp and Medusa.

Ans. Both are radially symmetrical.

Q.3. Obelia is—

(A) Asymmetrical. (B) Bilaterally symmetrical.
(C) Radially symmetrical. (D) Spherically symmetrical.

Ans. (C)

Q.4. In Obelia, the ectoderm is composed of—

(A) Two types of cells. (B) Three types of cells.
(C) Four types of cells. (D) Five types of cells.

Ans. (D)

Q.5. Radial symmetry is found in —

(A) Taenia. (B) Vaginulus.
(C) Hydra. (D) Physalia.

Ans. (D)

Q.6. Polyp helps in —

(A) Reproduction. (B) Nutrition.
(C) Excretion. (D) Respiration.

Ans. (B)

Q.7. Locomotory organ in Coelentera is—

(A) Setae. (B) Cilia.
(C) Flagella. (D) Basal disc or foot.

Ans. (D)

Q.8. Polyp is a/an —

(A) Reproductive zooid. (B) Nutritional Zooid.

(C) Asexually reproductive zooid. (D) None of the above.

Ans. (B)

Q.9. Medusa is a/an —

(A) Reproductive zooids. (B) Nutritive Zooid.

(C) Asexually reproducing (D) None of the above.

Ans. (A)

Q. 10. Polyps and Mesusa show —

(A) Asymmetry. (B) Bilateral symmetry.

(C) Radial symmetry. (D) None of the above

Ans. (C)

Q.11. An example of the order Hydroida is —

(A) Millepora. (B) Obelia.

(C) Stylaster. (D) Physalia.

Ans. (B)

Q.12. Which one of the following exhibits highest degree of polymorphism.

(A) Sycon. (B) Meandrina.

(C) Madrepoa. (D) Physalia.

Ans. (D)

Q. 13. Give an example of the class Anthozoa.

And. Tubipora.

PLATYHELMINTHES

Q.1. Describe the structure and life history of Taenia.

Ans. See pages 66 to 69 of the text.

Q.2. The body wall of Taenia consists of three layers. Which one of them is wrong.

(A) Cuticle. (B) Musculature.
(C) Mesenchyma. (D) Prosenchyme.

Ans. (D)

Q.3. Platyhelminthes are—

(A) Bilaterally symmetrical. (B) Radially symmetrical.
(C) Asymmetrical. (D) None of the above.

Ans. (A)

Q.4. In *Taenia solium*—

(A) the male reproductive organ consists of a single testis.

(B) the female reproductive organ has numerous ovaries.

(C) the female reproductive organ possesses mehlis glands.

(D) male and female reproductive organs are present at alternate segments.

Ans. (C)

Q.5. The stage which passes in the final host the sheep in the life history of *Fasciola hepatica* is—

(A) Fresh water snail. (B) Submerged vegetation.
(C) Pig. (D) Man.

Ans. (C)

Q.6. The bladder worm stage of Taenia is known as

(A) Rostellum. (B) Cysticercus.

(C) Onchosphere. (D) Strobila.

Ans. (B)

NEMATHELMINTHES

Q.1. Give the general characters of Phylum Nemathelminthes. Classify it giving examples.

Ans. See chapter 3 of the book.

Q.2. Which one of the following animals is sexually dimorphic ?

(A) Taenia. (B) Fasciola.

(C) Ascaris. (D) Pheretima.

Ans. (C)

Q.3. Cuticle of Ascaris is composed of—

(A) One layer of albuminous proteins.

(B) Nine layers of sucrose.

(C) Two layers of albuminous proteins.

(D) None of the above.

Ans. (A)

Q.4. Male ascaris is —

(A) Smaller in size than that of female.

(B) Larger in size than that of female Ascaris.

(C) Equal in size of the female Ascaris.

(D) None of the above.

Ans. (A)

Q.5. Infection of Ascaris usually occurs by—

(A) Tse-Tse fly. (B) Mosquito.

(C) House fly (D) Contamination of water and food.

Ans. (D)

Q.6. Post anal portion is —

(A) Curved in male ascaris.

(B) Curved in female Ascaris.

(C) Curved in male and female ascaris.

(D) Absent in female Ascaris.

Ans. (A)

Q.7. Ascaris is found in the —

(A) Stomach of man. (B) Small intestine of man.

(C) Large intestine of man. (D) None of the above.

Ans. (B)

ANNELIDA

Q.1. Body of leech is divisible into —

(A) 33 segments. (B) 35 segments.

(C) 43 segments. (D) 45 segments.

Ans. (A)

Q.2. In leeches—

(A) 2 pairs of eyes are present.

(B) 3 pairs of eyes are present.

(C) 4 pairs of eyes are present.

(D) 5 pairs of eyes are present.

Ans. (D)

Q.3. Two prominent suckers in leech are present on the —

(A) anterior side. (B) posterior side. (C) lateral side

(D) One on the anterior side and the other on the posterior side.

Ans. (A)

Q.4. Salivary glands in leech are —

(A) Unicellular and numerous.

(B) Unicellular and few.

(C) Large multicellular and unpaired.

(D) Small multicellular and paired.

Ans. (A)

Q.5. Parapodia are characteristics of —

(A) Amphipoda. (B) Diplopoda.

(C) Polychaeta. (D) Pauropoda.

Ans. (C)

Q.6. Enzyme hirudin is secreted by—

(A) Crop. (B) Salivary glands.

(C) Intestine. (D) Body wall.

Ans. (B)

Q.7. Excretory organs in Annelida are—

(A) Nephridia. (B) Coelomoducts.

(C) Flame cells. (D) Malpighian tubules.

Ans. (A)

Q. 8. In Nereis, the locomotory organs are —

(A) Setae. (B) parapodia.

(C) jointed appendages. (D) swimmerets.

Ans. (B)

ARTHROPODA

Q.1. Abdomen of –

(A) male cockroach is broader than the female cockroach,

(B) female cockroach is broader than male cockroach,

(C) male and female cockroach are of same size,

(D) None of the above.

Ans. (B)

Q.2. Anal styles are present in –

(A) Male cockroach, (B) Female cockroach,

(C) Both of the above, (D) None of the above.

Ans. (A)

Q.3. A broad porch is present in –

(A) Male cockroach, (B) Female cockroach,

(C) Both of the above, (D) None of the above.

Ans. (B)

Q.4. Cockroach is –

(A) Oviparous, (B) Viviparous,

(C) Both of the above, (D) None of the above.

Ans. (A)

Q.5. The mouth parts of cockroach are of –

(A) Biting type, (B) Chewing type,

(C) Both of the above, (D) None of the above.

Ans. (C)

Q.6. In male cockroach, the genital aperture lies between the –

(A) 9th and 10th sterna, (B) 9th and 8th sterna,

(C) 8th and 10th interna, (D) 7th and 9th sterna.

Ans. (C)

Q.7. The female genital aperture lies at the –

(A) 7th sternum, (B) 8th sternum,

(C) 9th sternum, (D) 10th sternum.

Ans. (B)

Q.8. In cockroach the respiratory organs are –

(A) Gills, (B) Lungs, (C) Trachea,

(D) Lungs and Trachea.

Ans. (C)

Q.9. Malpighian tubules are the excretory organs of–

(A) Cockroach, (B) Earthworm, (C) Rabbit,
(D) Rat.

Ans. (A)

Q.10. Compound eyes are present in –

(A) Earthworm, (B) Round worm,

(C) Tape worm, (D) Cockroach.

Ans. (D)

Q.11. Cockroach is –

(A) Unisexual, (B) Hermaphrodite,

(C) Neuter, (D) None of the above.

Ans. (A)

Q.12. Body of cockroach is divided into –

(A) Head and abdomen, (B) Head and thorax,

(C) Head, thorax and abdomen,

(D) Head, neck and thorax.

Ans. (C)

Q.13. What is the use of Malphigian tubule in Periplaneta ?

Ans. The Malpighian tubules excrete the nitrogenous wastes from the haemolymph of the haemocoel and empty the excreted substance into the gut. These are also osmoregulatory in function because they help in conserving a sufficient of water which has helped the insect in leading effective life activities in the terrestrial habitat.

MOLLUSCA

Q.1. Pila globosa is an example of Class –

(A) Gastropoda, (B) Pelecypoda,

(C) Scaphopoda, (D) Monoplacophora.

Ans. (A)

Q.2. Pila respires in water with the help of –

(A) Pulmonary sacs, (B) Filmentous gills,

(C) Ctenidium, (D) Osphradium.

Ans. (C)

Q.3. Ctenidium is an organ of –

(A) Digestive system of Pila,

(B) Respiratory system of Pila,

(C) Nervous system of Pila,

(D) None of these.

Ans. (B)

Q.4. Pulmonary sac is an organ of –

(A) Aquatic respiration of Pila,

(B) Aerial respiration of Pila,

(C) Both of these,

(D) None of these.

Ans. (B)

Q.5. Patella is an example of Phylum –

(A) Mollusca, (B) Arthropoda,

(C) Coelenterata, (D) Protozoa.

Ans. (A)

Q.6. Scaphopoda is a class of Phylum –

(A) Protozoa, (B) Coelenterata,

(C) Mollusca, (D) Echinodermata.

Ans. (C)

Q.7. Epitaenia plays a vital role during –

(A) Both aquatic and aerial respiration,

(B) Aerial respiration,

(C) Aquatic respiration

(D) None of these.

Ans. (A)

Q.8. Name the digestive gland found in Pila.

Ans. Liver or hepatopancrease.

Q.9. Name an example of Class Scaphopoda.

And. Dentalium.

Q.10. What is the function of Ctenidium ?

Ans. Ctenidium (gill) helps in aquatic respiration.

Q.11. Name the different kinds of sperms found in Pila.

Ans. Eupyrene and Oligopyrene.

ECHINODERMATA

Q.1. The Echinoderms are –

(A) Monoblastic, (B) Diplobastic,

(C) Triploblastic, (D) None of these.

Ans. (C)

Q.2. Asterias (Starfish) is –

(A) Asymmetrical,

(B) Radially symmetrical,

(C) Bilaterally symmetrical

(D) None of these.

Ans. (B)

Q.3. Water vascular system is the most characteristic feature of –

(A) Echinoderms, (B) Arthropods,

(C) Both of these, (D) None of these.

Ans. (A)

Q.4. Excretory system is absent in –

(A) Echinoderms, (B) Annelida,

(C) Arthropods, (D) None of these.

Ans. (A)

MINOR PHYLA

Q.1. Bugula avicularia belongs to –

(A) Brachiopoda, (B) Phoronida,

(C) Nemertinea, (D) Ectoprocta.

Ans. (D)

Q.2. Lingula belongs to –

(A) Nemertinea (B) Brachiopoda,

(C) Phoronida, (D) Ectoprocta.

Ans. (B)

Q.3. In which minor phylum does Phoronis belong ?

(A) Nemertinea, (B) Phoronida,

(C) Rotifera, (D) Bryozoa.

Ans. (B)

Q.4. Give an example of Rotifera.

Ans. Rotifera.

Q.5. An example of the minor phylum Nemertinea is

(A) Prostoma, (B) Pedalia,

(C) Phoronis, (D) Ligula.

Ans. (A)

Q.6. Give an example of Sipunculida.

Ans. Sipunculus.

****